GLOBAL AGRICULTURE: DEVELOPMENTS, ISSUES, AND RESEARCH

DAIRY FARMING

OPERATIONS MANAGEMENT, ANIMAL WELFARE AND MILK PRODUCTION

Global Agriculture: Developments, Issues, and Research

Additional books in this series can be found on Nova's website under the Series tab.

Additional e-books in this series can be found on Nova's website under the eBooks tab.

Agriculture Issues and Policies

Additional books in this series can be found on Nova's website under the Series tab.

Additional e-books in this series can be found on Nova's website under the eBooks tab.

GLOBAL AGRICULTURE: DEVELOPMENTS, ISSUES, AND RESEARCH

DAIRY FARMING

OPERATIONS MANAGEMENT, ANIMAL WELFARE AND MILK PRODUCTION

ANKE HERTZ
EDITOR

NOTICE TO THE READER

Library of Congress Cataloging-in-Publication Data

ISBN: 978-1-53613-969-3

Published by Nova Science Publishers, Inc. † New York

CONTENTS

PREFACE

According to the European legislation (Regulation EC No 853/2004), the food business operators collecting raw milk intended for the production of milk and dairy products must ensure compliance with certain health requirements for the animals. The animals must not show any symptoms of infectious diseases transmittable to humans, or signs of diseases of the udder or the genital tract that could contaminate milk. Furthermore, they must belong to a holding free or officially free of tuberculosis and brucellosis, and no unauthorized substances or authorized drugs must have been administered without respect to the withdrawal period. *Dairy Farming: Operations Management, Animal Welfare and Milk Production* presents a study with the goal of evaluating the compliance with the mentioned criteria in milk samples collected from 100 different dairy farms located in Central Italy. Additionally, under European milk quotas, dairy farms in Europe were limited in the amount of milk they could produce. While quotas were gradually increased over the past four decades, European milking quotas were completely abolished in April 2015 to help meet an expected 20% increase in the global consumption of milk and dairy products by 2050. With this, European dairy farmers can freely expand milk production based upon expected milk prices controlled by open market supply and demand. The authors present a review focused on milk production forecasting models and data variation from a past and future perspective. A comprehensive review of model applications and comparisons from studies over the past two decades is carried out, and both classical and modern methods are reviewed analysed. The concluding review focuses on scientific LCA studies conducted on a variety of different milk production systems, including the treatment of milk co-products, different allocation methods, the assessment of environmental impacts caused by fertilizers and agrochemicals in feed grain production and in the different stages of milk production. However, the methodology requires a higher degree of standardization, especially for the analysis of complex agricultural and livestock systems and their various forms and characteristics.

Chapter 1 - The milk from dairy animals has a good nutritional value because it represents an indispensable source of high quality proteins. Moreover, it is the raw material for many dairy products consumed worldwide such as fresh and ripened cheeses, butter, cream and fermented derivatives like yoghurt, kefir, koumiss, etc. According to the European legislation (Regulation EC No 853/2004), the food business operators collecting raw milk intended to the production of milk and dairy products must ensure compliance with some health requirements of the animals. More in detail, the animals must not show any symptoms of infectious diseases for humans, or signs of diseases of the udder or the genital tract that could contaminate milk. Furthermore, they must belong to a holding free or officially free of tuberculosis and brucellosis and no unauthorized substances or authorized drugs without respect of the withdrawal period must have been administered.

In addition to these requirements, raw milk coming from cows must meet some criteria with regards to the plate count at 30°C (ufc/ml) and somatic cell count (cells/ml). The bacterial contamination of milk can adversely affect its quality and safety and high values of somatic cell count can be considered an important indicator of mastitis. The results for these two regulatory parameters, i.e., the plate count at 30°C and somatic cell count, must be expressed as a rolling geometric average over a two- or three-month period, with at least two samples or one per month, respectively.

The aim of this study was the evaluation of compliance with the mentioned criteria in milk samples collected from 100 different dairy farms located in Central Italy. The plate count at 30°C was exceeded in 12 dairy farms while the somatic cell count just only in 2. These results demonstrated a good management of hygiene practices during milking and collection of raw milk. This approach could include some important elements such as the definition of primary udder health parameters, the detection of cows causing the problem and the implementation of a good herd management plan.

Chapter 2 - Under European milk quotas, dairy farms in Europe were limited in the amount of milk they could produce. While quotas were gradually increased over the past four decades, European milking quotas totally were abolished in April 2015, to help meet an expected 20% increase in the global consumption of milk and dairy products by 2050. With this, European dairy farmers can freely expand milk production based upon expected milk prices, now controlled by open market supply and demand. With increasing herd sizes and milk production, the ability to predict future production levels using empirical curve fitting prediction models developed on lower levels of milk production becomes problematic. This is due to increased milk production resulting in a larger peak to trough ratios. This problem is more acute in countries with primarily pasture-based dairy farming systems such as Ireland, where the total output of milk production closely follows the pattern of a regular lactation cycle, unlike countries with primarily stall based systems. Many curve fitting techniques rely on regression coefficients based upon the peak production of the lactation cycle. Thus, due to the increase in milk production over time, a classical fitting techniques could become less effective in the future. Therefore, there is a requirement for more modern adaptive techniques for milk production forecasting. This review focuses on milk production forecasting models and data variation from a past and future perspective. A comprehensive review of model applications and comparisons from studies over the past two decades is carried out. Both classical and modern methods are reviewed and their applications analysed.

Chapter 3 - Today, milk and dairy products are an important part of the human diet all over the world. The food processing sector is responsible for negative impacts on the environment, which increase with more intensive production. In this context, the growing interest in sustainable food production has called for the development of methods to increase productivity and simultainously maintain or even reduce the level of natural resources consumption. Life Cycle Assessment (LCA) is a method for the evaluation of environmental impacts of products during their life cycle, which permits the identification of critical points in processes and production stages and the description of their environmental as well as resource-related issues. This review focuses on scientific LCA studies conducted on a variety of different milk production systems, including the treatment of milk co-products, different allocation methods, the assessment of environmental impacts caused by fertilizers and agrochemicals in feed grain production and in the different stages of milk production. However, the methodology requires a higher degree of standardization, especially for the analysis of complex agricultural and livestock systems and their various forms and characteristics. Additional studies are necessary, for example on comparable dairy farm systems, in order to construct a wider database. The findings of those studies can help to improve resource use and productivity, and consequently, the environmental performance of the sector. The promotion of more sustainable practices is considered one of the most important contributions of LCA studies on dairy production systems.

In: Dairy Farming
Editor: Anke Hertz
ISBN: 978-1-53613-969-3

Chapter 1

MONITORING OF THE REGULATORY CRITERIA FOR RAW MILK COLLECTED FROM DAIRY FARMS IN CENTRAL ITALY

Maria Schirone[*] and Pierina Visciano
Faculty of Bioscience and Technology for Food,
Agriculture and Environment, University of Teramo, Teramo, Italy

ABSTRACT

The milk from dairy animals has a good nutritional value because it represents an indispensable source of high quality proteins. Moreover, it is the raw material for many dairy products consumed worldwide such as fresh and ripened cheeses, butter, cream and fermented derivatives like yoghurt, kefir, koumiss, etc. According to the European legislation (Regulation EC No 853/2004), the food business operators collecting raw milk intended to the production of milk and dairy products must ensure compliance with some health requirements of the animals. More in detail, the animals must not show any symptoms of infectious diseases for humans, or signs of diseases of the udder or the genital tract that could

[*] Corresponding Author E-mail: mschirone@unite.it.

contaminate milk. Furthermore, they must belong to a holding free or officially free of tuberculosis and brucellosis and no unauthorized substances or authorized drugs without respect of the withdrawal period must have been administered.

In addition to these requirements, raw milk coming from cows must meet some criteria with regards to the plate count at 30°C (ufc/ml) and somatic cell count (cells/ml). The bacterial contamination of milk can adversely affect its quality and safety and high values of somatic cell count can be considered an important indicator of mastitis. The results for these two regulatory parameters, i.e., the plate count at 30°C and somatic cell count, must be expressed as a rolling geometric average over a two- or three-month period, with at least two samples or one per month, respectively.

The aim of this study was the evaluation of compliance with the mentioned criteria in milk samples collected from 100 different dairy farms located in Central Italy. The plate count at 30°C was exceeded in 12 dairy farms while the somatic cell count just only in 2. These results demonstrated a good management of hygiene practices during milking and collection of raw milk. This approach could include some important elements such as the definition of primary udder health parameters, the detection of cows causing the problem and the implementation of a good herd management plan.

INTRODUCTION

Milk is the product of the mammary gland of many different animals such as cows, buffaloes, sheep, goats and donkeys. It is virtually sterile when secreted into the alveoli of the udder, but it can be contaminated during milking and/or handling with personnel, equipment and environmental sources (Sarkar, 2015). Due to its high nutrient content and low acidity, it represents an excellent growth medium for different bacteria, which can multiply based on temperature, other competing microorganisms and their metabolic products. Therefore, it is a very perishable commodity (Claeys et al., 2013; Ritota et al., 2017).

The microbiological quality of milk can be affected by the health status of dairy animals. In fact, they can show an inflammation of the mammary gland, as clinical or subclinical mastitis, determining the alteration of milk composition (Yu et al., 2017). Besides an animal welfare problem, mastitis can be considered also a food safety concern as well as one of the biggest economic damage sources for the herd management. It can be caused by a wide spectrum of pathogens (i.e., *Staphylococcus aureus*, *Streptococcus agalactiae*, *Corynebacterium bovis*, *Mycoplasma* spp.) spreading from animal to animal, primarily during milking. Their presence tend to result in chronic infections with flare-ups of clinical events (Abebe et al., 2016). This disease can be classified as clinical or subclinical. The usual signs of inflammation are present in the first form, characterized by red, hot and swollen mammary glands, while the subclinical mastitis is rarely diagnosed because no visible signs are seen, but generally milk production decreases and the somatic cell count (SCC) increases (Khan and Khan, 2006). The milk somatic cells include 75% leukocytes (i.e., neutrophils, macrophages, lymphocytes and erythrocytes), produced by the animals' immune system to fight the inflammation of the mammary gland, and 25% epithelial cells. The white blood cells can increase up to 99% in animals with mastitis (EFSA, 2015). The main factors affecting the SCC other than mastitis can be: i) stage of lactation; ii) age and breed of dairy animals; iii) season, stress and diurnal variation (Sharma et al., 2011). The increase of the SCC can cause the reduction in casein, fat and lactose of milk, affecting consequently the quality and yield of the derived dairy products (Malek dos Reis et al., 2013).

Another important indicator of milk quality is the total aerobic mesophilic bacteria count. It can be increased by several conditions, such as unhygienic practices during milking, dirty cow udders, teat injuries due to inadequate stall or platform design and late or insufficient milk cooling (Ürkek et al., 2017).

The European legislation (Regulation EC No 853/2004) set the criteria for raw cows' milk as follows:

- plate count at 30°C (ufc/ml) ≤ 100 000
- somatic cell count (cells/ml) ≤ 400 000

The value of the plate count at 30°C must be ≤ 1 500 000 ufc/ml for raw milk collected from other species, but if it is intended for the manufacture of products made by a process that does not involve any heat treatment, this value must be ≤ 500 000 ufc/ml.

In addition, the plate count at 30°C must be calculated as rolling geometric average over a two-month period, with at least two samples per month, whereas for the SCC the rolling geometric average over a three-month period, with at least one sample per month, must be considered. The geometric mean calculation generally produces a value somewhat less than the arithmetic mean for the same data set. This method has been chosen because a single high count in a data set has a greater impact on the arithmetic mean than the geometric mean. Moreover, only one very high value will not trigger regulatory action when using the geometric mean procedure (Sharma et al., 2011).

The aim of this chapter was the evaluation of compliance with the regulatory criteria in cows' raw milk samples collected from 100 different dairy farms, located in 5 regions of Central Italy during the year 2017.

MATERIALS AND METHODS

Raw milk samples were collected from 100 different dairy farms, located in the following regions of Central Italy: Abruzzo, Lazio, Marche, Molise, Umbria (Figure 1). The sampling was carried out during the first two quarters (from January to March and from April to June) of the year 2017. More in detail, the samples were collected twice a month and calculated as rolling geometric average over a two-month period for the plate count at 30°C, for a total of 1 200 samples in the first two quarters of the year, whereas they were collected one sample per month and calculated as rolling geometric average over a three-month period for the SCC, for a total of 600 samples.

All farms used automatic milking. The samples were put into 10 ml sterile tubes and immerged in icebox until arrival at the laboratory.

The analyses were performed according to Regulation EC No 2074/2005 following the reference methods: EN/ISO 4833 for the plate count at 30°C and ISO 13366-1 for the SCC.

Figure 1. Map of the sampling sites located in 5 regions of Central Italy.

The statistical analysis of the results was carried out using statistical software program (GraphPad Software Version 3.0 (Inc, San Diego, CA, USA). The ANOVA test was used to evaluate the effect of the collecting period.

RESULTS AND DISCUSSION

The data reported in this study resumed the rolling geometric average of the plate count at 30°C and the SCC in cows' raw milk samples collected from a total of 100 dairy farms, located in 5 regions of Central Italy. The results showed a general good hygienic condition of all dairy farms, indicating a suitable management system. The regulatory limits were exceeded in 12% of farms, independently from the investigated quarter of the year. The maximum value reached about 300 000 ufc/ml (Figure 2). Regarding to the SCC, only 2% of dairy farms exceeded the limit of 400 000 cells/ml up to about 700 000 cells/ml (Figure 3). The statistical analysis showed a significant difference ($p < 0.01$) between the two examined quarters of 2017 from all regions for the plate count at 30°C, and only between Abruzzo and Marche regions for the SCC.

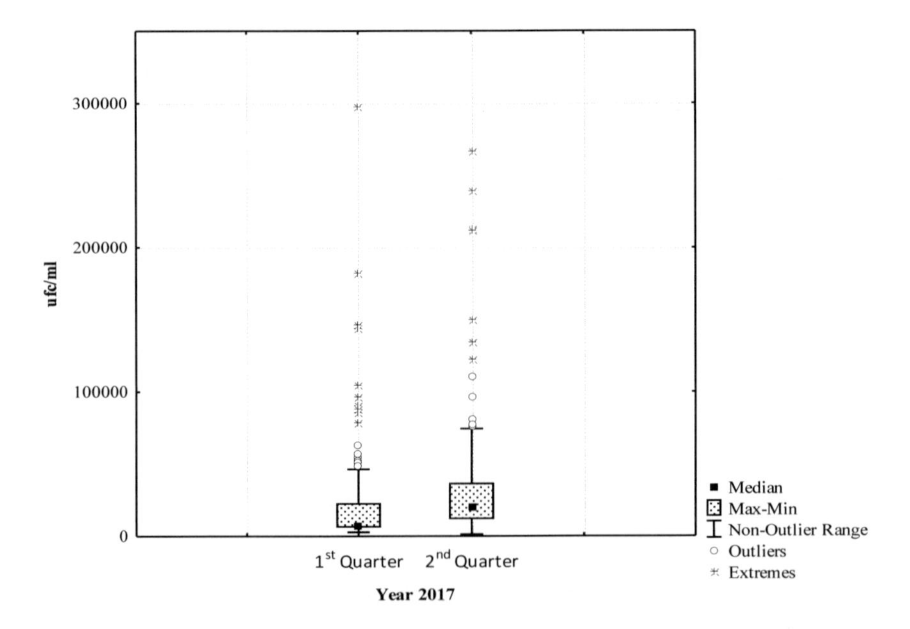

Figure 2. The plate count at 30°C related to the investigated dairy farms in the first two quarters of the year 2017.

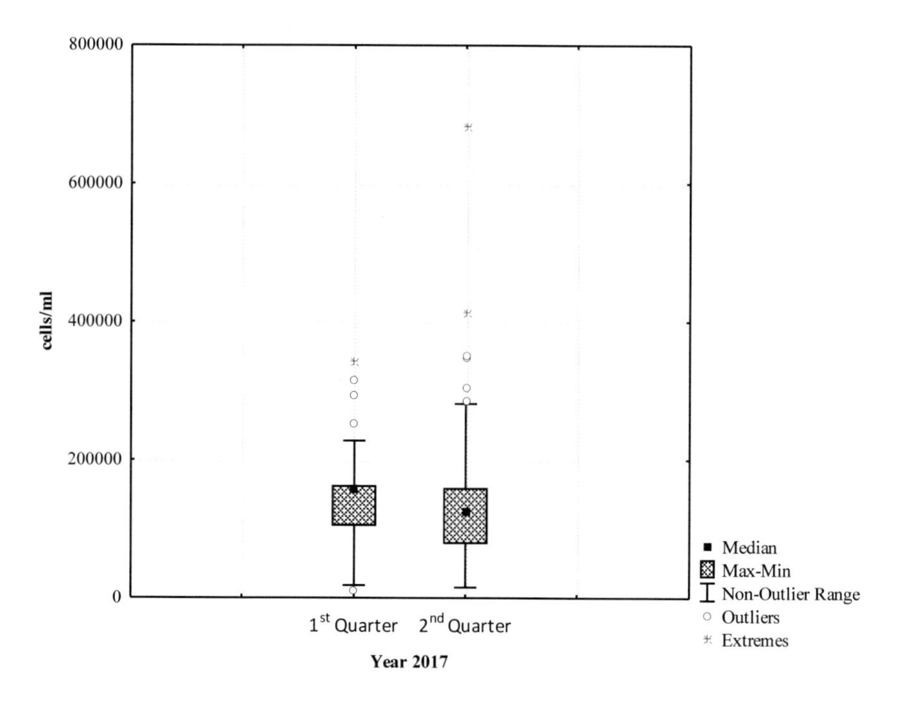

Figure 3. The somatic cell count related to the investigated dairy farms in the first two quarters of the year 2017.

This variability could be due to the different management of the investigated dairy farms, with regards above all to milking hygiene, such as wearing of gloves during milking and adopting a post-milking teat disinfection, but also to other measures referred to the treatment of mastitis during non-lactating period and the implementation of regular mastitis prevention (Abebe et al., 2016). The increase of the plate count at 30°C could be related to mastitis, environmental microorganisms, dirty milking equipment or failure of refrigeration (Kelly et al., 2009). The spread of microorganisms among cows can occur through vectors, such as milking machines, human hands or flies (Vanderhaeghen et al., 2015). Moreover, an indirect contamination may arise from the faecal matter of the individual itself, or other cows contaminating the udder and teats (EFSA, 2015). Also milk storage temperature and length, season of production, herd sizes, farm geographical location and hygiene practices can affect the microbial milk quality (Ramsahoi et al., 2011).

The microbial contamination of cows' udder and teats can be associated with the skin of animals, as well as the environment in which cows are housed and milked. So, pre-milking udder hygiene techniques could affect the plate count at 30°C of milk and a good cleaning of the teat with the sanitizing solution followed by drying with a clean towel is effective in reducing this value. In addition, the stress on the cow's teat produced by the action of milking machine can open the teat canal allowing the entry of bacteria capable of infecting the mammary glands (Ledenbach and Marshall, 2009). Also, the cleaning of milking apparatus can influence the total bacteria count. For instance, milk residue left on the contact surfaces of used equipment can favorite the growth of microorganisms.

Other authors (Bolzoni et al., 2015) reported that 29% out of 5 200 dairy farms exceeded the SCC limit one or more times a year, but they returned under this limit within 3 months; another 27% among the positive dairy farms remained non-compliant after this period. Bogdanovičová et al. (2016) evaluated the microbiological quality of cow's, goat's and sheep's milk in the Czech Republic collected from 41 dairy farms. The plate count at 30°C exceeded the regulatory limits in 13% of the samples, whereas the highest SCC value in cow's milk was 990 000 cells/ml.

According to Regulation EC No 853/2004, when raw milk fails to meet the legislative criteria, the food business operator must inform the competent authority and take measures to correct this situation. If these parameters do not return to be regular within the following three months, the delivery of raw milk from the production dairy farm will be suspended or subjected to requirements concerning its treatment and use necessary to protect public health (Regulation EC No 854/2004). Therefore, milk collection from the positive dairy farms of this study was excluded until these criteria returned to be regular. In that occasion, the farmers improved milking hygiene and in the meantime, they resorted to veterinary care for all the suspected ill cows.

Conclusion

The microbiological load of raw milk produced at dairy farm level can affect quality, shelf life and safety not only of this product itself, but also of the derived dairy products. The exceeding of the regulatory criteria for raw milk generally depends on animal's health status with regards to the SCC, even if the farmers are not able to recognize animals with mastitis. Instead, the high values of the plate count at 30°C can reflect poor hygienic conditions as a consequence of an improper management system. The application of sanitary practices on dairy farm, such as teat disinfection pre- and post-milking, correct handling, collection and storage of milk and staff hygiene represent important strategies to reduce the contamination of this commodity. In addition, the implementation of standard mastitis prevention and/or control programs can result in the reduction of prevalence and incidence of this disease in cows' herds.

References

Abebe, R., Hatiya, H., Abera, M., Megersa, B., and Asmare, K. 2016. Bovine mastitis: prevalence, risk factors and isolation of *Staphylococcus aureus* in dairy herds at Hawassa milk shed, South Ethiopia. *BMC Veterinary Research*, 12(270), 1-11.

Bogdanovičová, K., Vyletělová-Klimešová, M., Babák, V., Kalhotka, L., Koláčková, I., and Karpíšková, R. 2016. Microbiological quality of raw milk in the Czech Republic. *Czech Journal of Food Science*, 34(3), 189-196.

Bolzoni, G., Marcolini, A., and Buffoli, E. 2015. End of the derogations to Regulation (EC) 853/2004 for cow's milk in Italy. *Italian Journal of Food Science*, 27(1), 118-125.

Claeys, W. L., Cardoen, S., Daube, G., De Block, J., Dewettinck, K., Dierick, K., De Zutter, L., Huyghebaert, A., Imberechts, H., Thiange, P., Vandenplas, Y., and Herman, L. 2013. Raw or heated cow milk consumption: Review of risks and benefits. *Food Control*, 31, 251-262.

Commission Regulation (EC) No 2074/2005 of 5 December 2005 laying down implementing measures for certain products under Regulation (EC) No 853/2004 of the European Parliament and of the Council and for the organisation of official controls under Regulation (EC) No 854/2004 of the European Parliament and of the Council and Regulation (EC) No 882/2004 of the European Parliament and of the Council, derogating from Regulation (EC) No 852/2004 of the European Parliament and of the Council and amending Regulations (EC) No 853/2004 and (EC) No 854/2004. *Official Journal of the European Union*, L 338, 1-27.

EFSA BIOHAZ Panel (EFSA Panel on Biological Hazards) 2015. Scientific Opinion on the public health risk related to the consumption of raw drinking milk. *EFSA Journal*, 13(1), 3940, 1-95.

Kelly, P. T., O'Sullivan, K., Berry, D. P., More, S. J., Meaney, W. J., O'Callaghan, E. J., and O'Brien, B. 2009. Farm management factors associated with bulk tank total bacteria count in Irish dairy herds during 2006/07. *Irish Veterinary Journal*, 62(1), 36-42.

Khan, M. Z., and Khan, A. 2006. Basic facts of mastitis in dairy animals: A review. *Pakistan Veterinary Journal*, 26, 204-208.

Ledenbach, L. H., and Marshall, R. T. 2009. Microbiological spoilage of dairy products. In: *Compendium of the microbiological spoilage of foods*, W. H. Sperber and M. P. Doyle (Eds), Vol. XII, 41-67. Springer International Publishing.

Malek dos Reis, C. B., Barreiro, J. R., Mestieri, L., Porcionato, M. A. F., and Santos, M. V. 2013. Effect of somatic cell count and mastitis pathogens on milk composition in Gyr cows. *BMC Veterinary Research*, 9(67), 1-7.

Ramsahoi, L., Gao, A., Fabri, M., and Odumeru, J. A. 2011. Assessment of the application of an automated electronic milk analyzer for the enumeration of total bacteria in raw goat milk. *Journal of Dairy Science*, 94, 3279-3287.

Regulation (EC) No 853/2004 of the European Parliament and of the Council of 29 April 2004 laying down specific hygiene rules for food of animal origin. *Official Journal of the European Union*, L 226, 22-82.

Regulation (EC) No 854/2004 of the European Parliament and of the Council of 29 April 2004 laying down specific hygiene rules for food of animal origin. *Official Journal of the European Union*, L 226, 83-127.

Ritota, M., Di Costanzo, M. G., Mattera, M., and Manzi, P. 2017. New Trends for the Evaluation of Heat Treatments of Milk. *Journal of Analytical Methods in Chemistry*, 2017, 1-12.

Sarkar, S. 2015. Microbiological Considerations: Pasteurized Milk. *International Journal of Dairy Science*, 10(5), 206-218.

Sharma, N., Singh, N. K., and Bhadwal, M. S. 2011. Relationship of somatic cell count and mastitis: an overview. *Asian-Australasian Journal of Animal Sciences*, 24(3), 429-438.

Ürkek, B., Şengül, M., Erkaya, T., and Aksakal, V. 2017. Prevalence and comparing of some microbiological properties, somatic cell count and antibiotic residue of organic and conventional raw milk produced in turkey. *Korean Journal of Food Science of Animal Resources*, 37(2), 264-273.

Vanderhaeghen, W., Piepers, S., Leroy, F., Van Coillie, E., Haesebrouck, F. and De Vliegher, S. 2015. Identification, typing, ecology and epidemiology of coagulase negative staphylococci associated with ruminants. *The Veterinary Journal*, 203, 44-51.

Yu, J., Ren, Y., Xi, X., Haung, W. and Zhang, H. 2017. A novel lactobacilli-based teat disinfectant for improving bacterial communities in the milks of cow teats with subclinical mastitis. *Frontiers in Microbiology*, 8(1782), 1-12.

In: Dairy Farming
Editor: Anke Hertz

ISBN: 978-1-53613-969-3

Chapter 2

A REVIEW OF MILK PRODUCTION FORECASTING MODELS: PAST AND FUTURE METHODS

Fan Zhang, Philip Shine, John Upton, Laurance Shaloo and Michael D. Murphy*

Department of Process, Energy and Transport Engineering,
Cork Institute of Technology, Co. Cork, Ireland

ABSTRACT

Under European milk quotas, dairy farms in Europe were limited in the amount of milk they could produce. While quotas were gradually increased over the past four decades, European milking quotas totally were abolished in April 2015, to help meet an expected 20% increase in the global consumption of milk and dairy products by 2050. With this, European dairy farmers can freely expand milk production based upon expected milk prices, now controlled by open market supply and demand. With increasing herd sizes and milk production, the ability to predict future production levels using empirical curve fitting prediction models developed on lower levels of milk production becomes problematic. This is due to increased milk production resulting in a larger peak to trough

* Corresponding Author Email: MichaelD.Murphy@cit.ie.

ratios. This problem is more acute in countries with primarily pasture-based dairy farming systems such as Ireland, where the total output of milk production closely follows the pattern of a regular lactation cycle, unlike countries with primarily stall based systems. Many curve fitting techniques rely on regression coefficients based upon the peak production of the lactation cycle. Thus, due to the increase in milk production over time, a classical fitting techniques could become less effective in the future. Therefore, there is a requirement for more modern adaptive techniques for milk production forecasting. This chapter focuses on milk production forecasting models and data variation from a past and future perspective. A comprehensive review of model applications and comparisons from studies over the past two decades is carried out. Both classical and modern methods are reviewed and their applications analysed.

Keywords: milk production forecasting, dairy modelling, lactation curve modeling

INTRODUCTION

With the abolishment of the European Union (EU) milk quotas in April 2015, dairy farmers were able to freely increase milk production. As a result, Irish annual milk yield increased by 13.3% between 2014 and 2015 (Table 2 and Figure 1). Due to the grass-based dairy system, the monthly milk yield varies seasonally throughout the year (as shown in Figure 1), as cows are kept indoors during the winter months. From 2010 to 2015, annual milk production in Ireland increased by almost 25%, on the other hand, the Peak to Trough Ratio (PTR) has increased from 4.7 to 5.7, while the UK has a relatively stable PTR of 1.2 over the same period (Central Statistics Office, 2017).

In addition to macro-level milk production figures, the average annual performance level (litres/cow) varied by between -4% and 5% in Ireland, as shown in Table 3 (Teagasc, 2011). Additionally, milk price variances and other feed and farm management related costs have considerable effects on net margin. By 2016, there were approximately 15,639 Irish dairy farms with an average income of €51,809, according to Teagasc

National Farm Survey Results (Teagasc, 2016). The average income of dairy farms has declined continuously in 2015 and 2016, as shown in Table 4, due to the reduction in milk price and gross output (Teagasc, 2016). Despite this reduction in the average dairy farm income, dairy farms still have the opportunity to recover by practicing positive technical and financial methods, including expanding milk production, increasing system efficiency on farms and reducing total costs.

Milk production within the EU has been restricted since the introduction of milk quotas in 1984 under the Dairy Produce Quota Regulations 1984. After three decades, the EU milk quota system was abolished in April 2015. As a free market, milk yield and price fluctuations pose a logistical challenge for both milk producers (farmers) and processors (creameries). The reporting of milk production statistics frequently use averaged values derived from cumulative milk yield over a relative fixed period (i.e., monthly, annual) at the herd level. According to the Food Wise 2025 report (Department of Agriculture Food & the Marine, 2016) and the Dairy Road Maps (Teagasc 2008, 2013, 2016), total number of dairy cows, average milk yield and net margin are the three major indicators of economic forecasting and comparison, as well as reference points of targets and achievements on the Dairy Road Maps (as shown in Table 5). This can be seen to result in a disjunction between targets (estimated values) from regulators and achievements (actual values) from industry. For example, differences between achievements and targets of average milk yield and net margin on all three Road Maps indicates that targets were far from fulfilled: even the achievements in 2016 were still below the targets in the Road Map 2018 which was made in 2008 (as shown in Figure 2). In a future scenario, a situation may arise whereby milk yield cannot be predicted precisely, potentially leading to overcapacity, under capacity and/or milk price volatility. With this, both milk producers and processors may benefit from accurate milk production information via practical forecasting methods. Accurate milk production forecasts would allow farmers to predict on farm thermal cooling loads, plant capacity sizing, plant operations and optimization (Breen et al., 2015; Murphy et al., 2015, 2014, Upton et al., 2015, 2014).

Table 1. Food and beverage exports of Ireland between 2014 and 2015 (Data source: Central Statistics Office)

	2014	2015	Rank	Change	Share of Total
Category	€M	€M		%	%
Dairy Products	3,105	3240	1	4%	30%
Beef	2,280	2,410	2	6%	22%
Others	5,085	5,157	-	-	48%
Total	10,470	10,825	-	3%	100%

Table 2. Comparison of milk production yield and trends between 2014 and 2015 in different countries (Data source: Central Statistics Office)

Milk Production (thousand tonnes)	Ireland	Rest of EU	Total EU	USA	New Zealand	Australia
2014	5,816	142,602	148,418	93,460	21,843	9,513
2015	6,589	145,043	151,632	94,571	21,533	9,605
% change	13.30%	1.70%	2.20%	1.20%	-1.40%	1.00%

Table 3. Average milk production yield (litres per cow), milk price (cent per litre) and net margin (cent per litre) of Irish dairy industry and annual changes (%) (2011-2016) (Data source: Teagasc)

	2011	2012	2013	2014	2015	2016
Average Yield						
Production (litres/cow)	5,166	4,968	5,135	5,170	5384	5646
Change (%)	0	-4%	3%	1%	5%	5%
Average Price and Margin						
Milk Price (cent/litre)	35.3	32.3	39.6	39.5	30.9	27.9
Change (%)	0	-9%	23%	0	-22%	-10%
Net Margin (cent/litre)	12.9	7.7	12.1	12.9	9.9	6.7
Change (%)	0	-41%	58%	7%	-24%	32%

On the other hand, regarding precision agriculture and animal welfare, precise forecasts of milk yield for a specific cow at the individual cow level could be beneficial within the dairy industry. Such beneficial applications include: 1) monitoring disease and the condition of a cow's health, i.e., mastitis detection (Andersen et al., 2011), conception interval prediction (Madouasse et al., 2010). 2) cow milking performance prediction (Nielsen et al., 2010; Rémond et al., 1997), decision support for advanced milking parlours and milking machines (André et al., 2010; Thomas and DeLorenzo, 1994). 3) precision input for herd simulation models (Petek and Dikmen, 2006; Ruelle et al., 2016). These applications will have direct or indirect effects on the milk yield of an individual cow. Consequently, as time goes on, the performance of the herd milk yield can be improved, as well as the prediction accuracy and precision of both herd and individual cow milk yield.

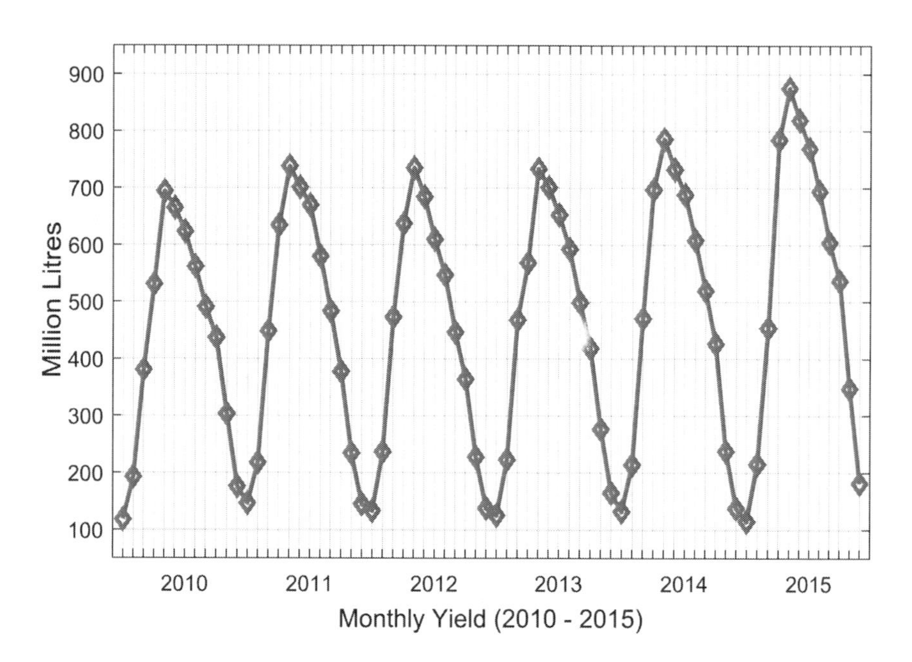

Figure 1. Monthly yield of Irish milk production from January 2010 to December 2015 (Data source: Central Statistics Office).

Table 4. Income of Irish dairy farms (2014-2016): total number of dairy farms, average gross output (Euro), average total costs (Euro), average income (Euro) and annual changes (%) (2014-2016) (Data source: Teagasc)

	2014	2015	2016
No. of Dairy farms	17,000	15,588	15,639
Gross Output (€)	185,685	180,115	168,339
Change (%)	0	-3%	-7%
Total Costs (€)	120,000	117,974	116,590
Change (%)	0	-2%	-1%
Average Income(€)	66,107	62,141	51,809
Change (%)	0	-6%	-17%

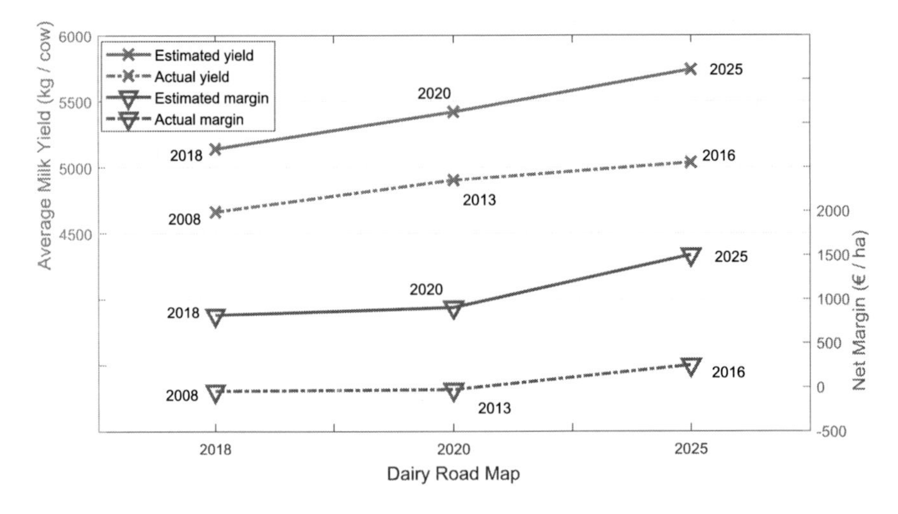

Figure 2. Comparison of estimated and actual values (annual average milk yield and net margin) on three Dairy Road Maps (2018, 2020 and 2025) (Data source: Teagasc).

With milk production levels expected to increase to 7.5 billion litres by 2020 (from a base of 4.9 billion litres in 2008-09), along with an increase in the national average herd size, accurate milk production forecasts would allow individual farmers to predict increased on-farm thermal cooling loads and to optimize the sizing and configurations of plant infrastructure (Department of Agriculture Food & the Marine, 2016). Concurrently, accurate milk production forecasts will be useful for farm management support and analysis for herd management, energy utilization and economic prediction (Shalloo et al., 2011, 2004; Murphy et al., 2013; Upton et al., 2015).

Table 5. Summary of Irish Dairy Road Maps (2018, 2020 and 2025): total dairy cows, annual average milk yield (kg/cow), net margin (€/ha) are three major indicators of economic forecasting on the Dairy Road Maps (Data source: Teagasc and Central Statistics Office)

	Key Figures of Dairy Road Map					
	Road Map 2018		Road Map 2020		Road Map 2025	
	Current in 2008	Targets in 2018	Current in 2013	Targets in 2020	Current in 2016	Targets in 2025
Dairy Farm Numbers	_	15,500	_	16,500	_	16,500
Dairy Cows Numbers (million) (by December)	1.104	1.382	1.082	1.395	1.295	1.7
Average Herd Size	_	89	_	85	_	>100
National Milk Production (million tonnes)	_	7,101	_	7,648	_	9,687
Average Milk Delivered per Farm (kg)	_	458,123	_	463,500	_	587,100
Milk Yield (kg/cow)	4,661	5,140	4,902	5,420	5,036	5,739
Net Margin (€/ha)	-42	821	-25	909	250	1,503

Modelling Milk Yield

Historically, studies have been undertaken regarding milk production prediction techniques where diverse equations have been developed for the purpose of describing a lactation curve based on past milk yield data. These equations include curve fitting models, regression models and auto-regressive models and mechanistic models. Each one of these models has been successfully applied to predict cow/herd level milk production, based on specific datasets. Each statistical model offers separate advantages related to their ease of deployment and ability to effectively quantify the non-linear nature of the lactation curve.

In relation to curve fitting models algebraic equations are utilised for fitting lactation curves using empirical data, usually requiring one variable as input data, such as daily or weekly cumulative milk yield at herd or individual cow level. Curve fitting models have performed well in numerous studies, taking many different forms including parabolic exponential (Sikka, 1950), incomplete gamma (Wood, 1967), polynomial (Ali and Schaeffer, 1987), exponential (Wilmink, 1987), Cubic splines (Green and Silverman, 1993), Legendre polynomial (Kirkpatrick et al., 1994) and log-quadratic (Adediran et al., 2012). Due to the variety of mathematical functions available to model lactation profiles, curve fitting models have two sub-categories: 1) empirical models (linear or nonlinear) and 2) semiparametric models which show their flexibility in fitting time-series for events with various curves (Schaeffer, 2004; Sherchand et al., 1995). However, lack of flexibility and adaptation is a common weakness of curve fitting models when dealing with significant fluctuations in yield within and between years (Jones, 1997).

Regression models have been found to perform well statistically over a wide variety of milk yield datasets. Auto-regressive neural network models were introduced for milk yield prediction and found to more accurately forecast milk yield when compared with static neural network models (Murphy et al., 2014). Mechanistic models offer more biological details and take account milk yield models, paddock models and grass conditions (Ruelle et al., 2016).

Empirical Algebraic Models

Empirical algebraic models have been utilised in research to forecast milk yield over the course of the lactation cycle (lactation curve) since early 20th century. Numerous studies have attempted to describe lactation curves using algebraic formula, these include Brody et al. 1923; Sikka 1950; Wood 1967; Wilmink 1987; Ali and Schaeffer 1987; Guo and Swalve 1995; Noreen Quinn 2005.

The curve-fitting model developed by Brody et al. (1923) was the first gamma model to forecast milk yield over the lactation cycle for four breed of cows in the US including, Holstein, Jersey, Guernsey and Scrub. In Brody et al.'s original formula:

$$Y_n = ae^{-bn} \tag{1}$$

Where Y_n is the milk yield during the nth month, a is the theoretical value of the milk yield at the time of parturition, and b is coefficient. The initial aim of this model was to describe the declining phases of the whole lactation. Hence, this model first proposed the concept of a constant relative rate of decline in milk yield of b kg per month from an initial value of a.

One year later, Brody et al. (1924) proposed a more complex model which uses two exponential functions to describe not only the declining phases but also the whole lactation for 119 US Holstein-Friesian cows. The model of Brody et al. (1924) was the first gamma model on the prediction of whole lactation yield research, the formula of which equalled:

$$Y_n = ae^{-bn} - ae^{-cn} \tag{2}$$

Where Y_n is the milk produced during the nth month, a is the theoretical value of the milk yield at the time of parturition, and b and c are coefficients. However, although this model was the state-of-art in that era, this model was later found to have underestimated in the mid lactation and overestimated in the late lactation.

A study investigating the effect of heredity and environment factors on milk yield was conducted by Sikka, L. C.(1950). The data used this study was obtained from five herds of Ayrshire cows and involved 2392 lactations during the period 1920 - 1939. Through utilising the multiple regression technique, Sikka concluded that any given lactation can be predicted much more accurately by using the following formula:

$$Y_n = ae^{(bn-cn^2)} \tag{3}$$

Where Y_n is milk production during the nth month, a is the theoretical value of the milk yield at the time of parturition, and b and c are coefficients. The model developed by Sikka et al. was found to have a better performance for the first lactation compared with the predictions of latter lactations due to the symmetric estimated yield around the peak yield.

Wood's model (1967) is the most commonly used model to predict milk yield throughout the whole lactation cycle and has been used as the base for consequent studies involving empirical equations of lactation curves. Wood's model was the first equation that presented the lactation curve in a reasonable accuracy using the following equation:

$$Y_n = an^b e^{-cn} \tag{4}$$

Where Y_n is the average daily yield in the *n*th week, a is a scaling factor associated with the average yield, and b and c are related to pre-peak curvature and post-peak curvature, respectively. Wood's model utilised the least squares method to get its regression parameters a, b and c. According to the prediction of Wood's model, the peak yield of $a(b/c)^b e^{-b}$ will occur at the (b/c)th week. However Wood's model was inherently non-linear and it was computationally expensive to perform nonlinear regression in 1960's.

A logarithmic transformation of Wood's model was a more popular technique, which makes the model's linear equation as follows (Equation 5 or Equation 6):

$$\log_e(Y_n) = \log_e(a) + b\log_e(n) - cn \tag{5}$$

$$\ln(Y_n) = \ln(a) + b\ln(n) - cn \tag{6}$$

In certain circumstances, a lack of fit was found in the predictions developed by Wood's model. As a consequence, Wilmink (1987) proposed an non-linear exponential model to predict milk yield with four parameters. In Wilmink's study, test-day records of 14,275 purebred Dutch Friesians were analysed by generalized least squares. In Wilmink's model:

$$Y_n = a + bn + ce^{-dn} \tag{7}$$

Where Y_n is the yield in lactation day n, where a, b, c and d are coefficients. The exponential term tends to zero as n increases, while after the peak, the decline in yield eventually equates the straight line a + bn. Although Wilmink (1987) claimed this non-linear model offered a greater representation of the lactation curve, other studies reported that a reduced d parameter value offered a simplified model with a similar level of forecasting accuracy (Olori et al., 1999; Brotherstone and White, 2000).

Concurrently, Ali and Schaeffer (1987) proposed the first polynomial regression model based on empirical data from 775 Canadian Holstein-Friesian cows in 42 herds (1964-1984), the formula of which is as follows:

$$Y_n = a + b\gamma + c\gamma^2 + d\omega + e\omega^2 + f \tag{8}$$

Where Y_n is the milk yield in lactation day n, $\gamma = n/305$, $\omega = \ln(305/n)$, f is the residual error and a, b, c, d, e are regression coefficients. In this formula, a is associated with peak yield, b and c are associated with the decreasing slope of the curve, d and e are associated with the increasing slope, and f is the residual error for this model. Ali and Schaeffer's model requires test data to estimate five parameters which is a disadvantage for some applications. Thus, this model could only be applied on milk yield data from a limited length lactation, and it was not suitable for extending part of the lactation. Moreover, the concave shape of Ali and Schaeffer's formula resulted in limitations as it could only be applied on milk yield forecast. The Ali and Schaeffer model has shown to be one of the most effective milk yield predictors over the last 30 years. A recent study found that the Ali and Schaeffer model performed better on the highly heterogeneous data, in contrast to the Wilmink model (Melzer et al., 2017).

Based on the Ali and Schaeffer model, Quinn et al. (2005) proposed the Ali-B model which have showed better forecasting performance than the original Ali and Schaeffer model based on data of 4336 Irish dairy cows from 79 spring-calving herds. After removing parameter b, the modified Ali and Schaeffer's model (the Ali-B model) was the most accurate model for predicting total and weekly milk yield. The Ali-B formula is as follows:

$$Y_n = a + c\gamma^2 + d\omega + e\omega^2 + f \tag{9}$$

Where Y_n is the daily milk yield in lactation day n, $\gamma = 7n/305$, $\omega = \ln(305/7n)$, f is the residual error and a, b, c, d, e are the regression coefficients. Peak yield is associated with the coefficient a, b and c are associated with the decreasing slope, d and e are associated with the increasing slope of the curve.

Guo and Swalve (1995) proposed a mixed logarithmic model, the formula equalling:

$$Y_n = a + b\sqrt{n} + c\ln(n) \tag{10}$$

Where, n is the number of weeks in lactation, and a and b are coefficients. Quinn et al. (2005) analysed the prediction performance of this model to those models developed by Wood (1967), Wilmink (1987), Ali and Schaeffer (1987) and Ali-B (2005) and found that the Ali-B model was the most consistent at satisfying the assumptions and prediction of weekly and total lactation individual milk yield.

Adediran et al. (2012) proposed Log-quadratic model for Australian pasture-based dairy systems (Equation 11). The data used including 9,505 lactations from 154 Holstein-Friesian herds collected from 2005-2007. This recent log-quadratic model has the peculiar ability to fit both inclining and declining lactation rates according Adediran et al.'s research results. The author stated that the developed model performed well for both the average lactation and individual cow lactations. However, the tested data of individual cows was the average of a selected group of cows due to the actual diversity of the individual cows. This limitation is ubiquitous for all empirical algebraic lactation models.

$$Y_n = \exp^{[a(b-\log n)^2+c]} \tag{11}$$

Semiparametric Approach

Recently, semiparametric functions including Legendre polynomial and Cubic spline have been applied for lactation curve modelling due to their flexibility in fitting time-series for events with various curves. Kirkpatrick et al. (1994) created the Legendre polynomial model which are n^{th} degree polynomial functions. The equation describing a single observation equals:

$$Y_n = \sum_{i=0}^{n} \alpha_i \varphi_i(\omega) \tag{12}$$

$$\omega = 2\left(\frac{t - t_{min}}{t_{max} - t_{min}}\right) - 1 \tag{13}$$

Where ω is lactation time unit ranging from -1 to +1, t is the test day, t_{min} (5 day) is the earliest days in milk (DIM) and t_{max} (305 day) is the latest DIM (Schaeffer, 2004).

$$\varphi_i(\omega) = \sqrt{\frac{2n+1}{2}}\, P_n(\omega) \tag{14}$$

Where $P_n(\omega)$ is a polynomial of degree n and $\varphi_i(\omega)$ is the normalized polynomial. The first 5 Legendre polynomials functions of standardized units of time (ω) are defined below, according to Spiegel (1971).

$$P_0 = 1$$

$$P_1(\omega) = \omega$$

$$P_2(\omega) = \frac{1}{2}(3\,\omega^2 - 1)$$

$$P_3(\omega) = \frac{1}{2}(5\,\omega^3 - 3\,\omega)$$

$$P_4(\omega) = \frac{1}{8}(35\,\omega^4 - 30\,\omega^2 + 3) \tag{15}$$

Different normalized Legendre polynomial functions of standardized units of time (ω) and coefficients α with different degrees require different data from observations by lactation. For example, degree 2, 3, 4 requires a minimum 4, 5, 6 observations by lactation respectively, which implies that they were not applicable to all the data of sampling groups. According the study from Silvestre et al. (2006), the Legendre polynomial functions with different degrees generated totally different accuracy results. Particularly, the Legendre polynomial functions were more accurate for describing the lactation curve when the first test day was recorded late in lactation than models developed by Wood, Wilmink and Ali and Schaeffer. These results supported the authors' hypothesis that the performances of Wood, Wilmink and Ali and Schaeffer models were greatly affected by both the sample properties and sample dimension.

The second semiparametric model is the Cubic spline fitting which was proposed by Green and Silverman (1994). The original normal formula is:

$$Y_k(x) = \sum_{i=0}^{k} \frac{\alpha_i x^i}{i!} + \sum_{j=1}^{n-1} \frac{\beta_i (x - x_j)^k}{k!} \quad (16)$$

Where

$$(x - x_j)^k = \begin{cases} (x - x_j)^k, & x \geq x_j \\ 0, x < x_j \end{cases}$$

Recently, the Cubic spline model was used to describe the lactation curve (Silvestre et al., 2005; White et al., 1999). The Cubic spline model requires a minimum of three observations for each record, and the formula for each record can be written as:

$$Y_n = a_i + b_i (n - n_i) + c_i (n - n_i)^2 + d_i (n - n_i)^3,$$
$$\text{for } n_i < n < n_{i+1} \quad (17)$$

According to other studies, the Cubic spline model fitted the lactation data best and had the additional advantage of describing the lactation curve adequately with fewer observations than was required for the Legendre polynomial model (Adediran et al., 2012; Silvestre et al., 2006). Although, semiparametric functions including Legendre polynomial and Cubic spline are not always the most accurate, these functions were found to work well using an appropriate data set as a previous study implied that a single outlier data point could greatly distort the curve, in particular for small data sets with few data points where the impact of outlier data is enhanced (Motulsky and Ransnas, 1987).

Surface Fitting Model

The surface fitting method creates a surface fit to the data in the x, y, and z planes. In this study, three training data matrices which were deemed the most accessible data for commercial dairy farms including number of cows milked (NCM), days in milk (DIM) and daily herd milk yield (DHMY) were chosen as the input datasets. In this study, the expression of the surface fitting method can be written as:

$$Z_{(x,y)} = \varepsilon + p_1x + p_2y + p_3x^2 + p_4xy + p_5y^2 + p_6x^3 + p_7x^2y + p_8xy^2 + p_9y^3 \quad (18)$$

Where $Z_{(x,y)}$ is the DHMY and the dependent variable, x is the independent variable DIM and y is the independent variable NCM, p_1, p_2, p_3, p_4, p_5, p_6, p_7, p_8, and p_9 are the surface coefficients and ε is the residual error.

Multiple Linear Regression

The Multiple Linear Regression (MLR) model has been proposed by multiple authors in cognate studies (Grzesiak et al., 2003; Kerr et al., 1998; Zar, 1984). The MLR model was chosen for milk yield prediction for two reasons. Firstly, the MLR model was proved to be successful in milk yield forecasting at the herd level (Dongre et al., 2012; Grzesiak et al., 2003; Sharma et al., 2007; Smith, 1968). Secondly, the MLR model can use more input variables than the curve fitting models, which can only use DIM and DHMY. Research carried out by Smith (Smith, 1968) has successfully demonstrated that the addition of rainfall and temperature data as additional input variables can improve the annual milk yield forecasting accuracy of a MLR model. For the purpose of forecasting herd level milk yield using most accessible data from commercial farms, Murphy et al. (2014) utilised a practical expression of the MLR model which only takes two inputs:

$$Y_n = \varepsilon + \alpha_1 NCM_n + \alpha_2 DIM_n \quad (19)$$

Where Y_n is the daily herd milk yield and the dependent variable, NCM and DIM are independent variables, α_1, and α_2 are the regression coefficients and ε is the residual error.

ANN Modelling Approach

An artificial neural networks (ANN) is a computational model based on the operating principles of the human nervous system and brain. An ANN is configured and applied to specific applications, such as function approximation, including non-linear function fitting (Esen et al., 2008; Kalogirou and Bojic, 2000; Specht, 1991), classification, including pattern recognition (Fukushima, 1988; Lyons et al., 2004), numerical control applications (Jung and Kim, 2007; Kim and Lewis, 2000) and time series prediction. Previous studies have applied ANN modelling in the forecasting domain (Hocaoĝlu et al., 2007; Kalogirou and Bojic, 2000; Khoshnevisan et al., 2013; Kim et al., 2004; Pahlavan et al., 2012; Voyant et al., 2011; Wong et al., 2010; Zarzalejo et al., 2005).

In this section, the ANN architectures and related algorithms are reviewed. The basic component of a neural network is a node (or neuron), which is designed to mimic the understanding of the functionality of a neuron in the human brain. Each node forms the basic block of a neural networks. Figure 3 shows the structure of a neuron node.

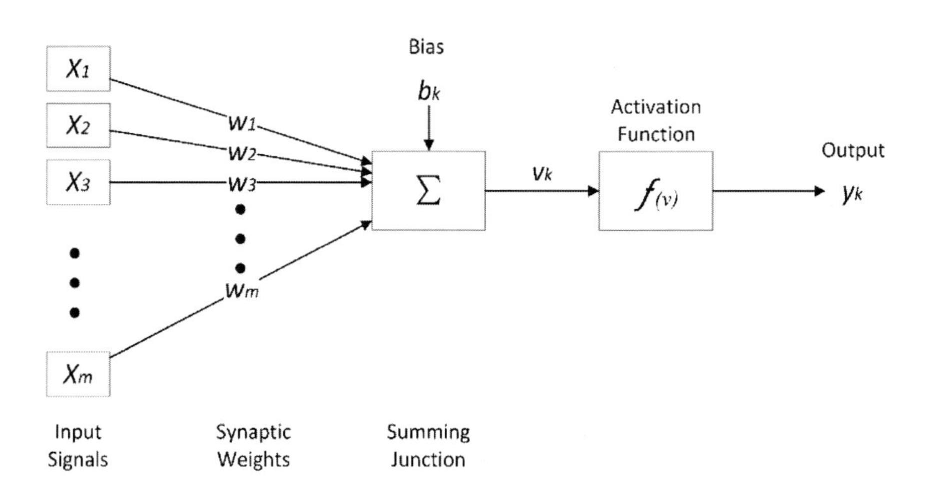

Figure 3. The structure of a neuron node.

The calculation flow of neuron node is as follows:

- The inputs x_j to a node are the measurements or the outputs from other nodes. Each input could be treated as a connection or a link with synaptic weights w_{kj}.
- Each node is characterized by an additive threshold value b_k and an activation function f (v). The threshold is used as an offset.
- The node sums the weighted inputs and the threshold value and passes the result through its characteristic nonlinearity to produce the output y_k

The summation of the weighted input signals is described as:

$$v_k = \sum_{j=1}^{m} w_{kj} x_j$$

Where x_j are input vectors, w_{kj} are the synaptic weights of neuron k, v_k is the linear summation of the weighted input vectors.

This allows the neuron node output to be written as:

$$y_k = f(v_k + b_k) \tag{20}$$

Where v_k is the linear summation of the weighted input vectors, b_k is the bias, f (v) is the activation function and y_k is the neuron output.

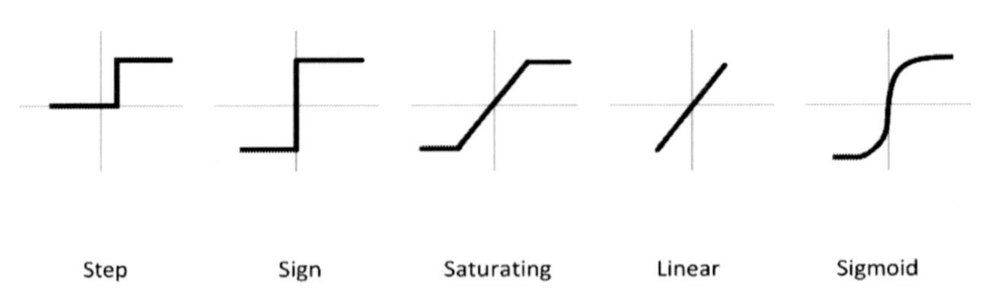

Figure 4. Typical activation functions.

The common types of activation functions include step, sign, saturating, linear, and sigmoid etc. (Figure 4).

- Step function:

$$f(x) = \begin{cases} 1, x \geq t \\ 0, x < t \end{cases}$$

- Sign function:

$$f(x) = \begin{cases} 1, \geq 0 \\ -1, x < 0 \end{cases}$$

- Saturating function:

$$f(x) = \begin{cases} 1, x > t \\ t, \ -t \leq x \leq t \\ -1, x < -t \end{cases}$$

- Linear function:

$$f(x) = t$$

- Sigmoid function:

$$f(x) = \begin{cases} = \frac{1-e^{-x}}{1+e^{-x}}, \ for \ -1 \leq f(x) \leq 1 \\ = \frac{1}{1+e^{-x}}, \ for \ 0 \leq f(x) \leq 1 \end{cases}$$

A typical neural networks usually consists of a three-layer architecture including an input layer, a hidden layer and an output layer (as shown in Figure 5). Input layer nodes and output layer nodes are accessible from the external environment. However, the hidden layer nodes are not directly accessible from the external environment, meaning that all input and output connections of these nodes are associated with nodes within the ANN only (black box). In addition, the external inputs to the network are usually not weighted while all interconnections within the network are weighted. There are two basic architectures for ANN: 1) the feedforward architectures and 2) the feedback architectures. Figure 5 demonstrates a typical static feedforward ANN. This type of ANN is commonly referred to as a Multilayer Neural Networks (MNN). The signals between the nodes of the feedforward ANN only flow in the forward direction. Nodes of a layer could have inputs from nodes of any of the earlier layers.

In the feedback ANN (also named recurrent ANN), the output signal from a node is allowed to flow in the forward and backward directions, potentially feeding back as an input to the same node itself in the input layer.

In comparison with other modelling approaches discussed in previous sections, the advantage of the ANN model is that neural networks can be trained by supervised learning methods and error updating rules. Hence, the ANN prediction performance can be improved due to synaptic weights adjustment and better outputs selection.

A number of studies have reported that the ANN technique can be utilized for milk yield forecasting (Dongre et al., 2012; Gorgulu, 2012; Grzesiak et al., 2006, 2003; Ince and Sofu, 2013; Khazaei and Nikosiar, 2005; Kominakis et al., 2002; Salehi et al., 1998; Sanzogni and Kerr, 2001; Sharma et al., 2007; Torres et al., 2005). However, the common disadvantage of these proposed milk prediction ANN models is the requirement of a large amount of detailed information for model inputs. One model developed by Sharma et al. requires 12 individual traits of each cow (genetic group, DMY, season of birth, period of birth, birth weight, age at maturity, weight at maturity, season of calving, period of calving, age at calving, weight at calving, peak yield, days to attain peak yield). Similarly, An ANN model developed by Lacroix et al. (1995) requires 16 parameters for input to the model (logarithm of somatic cell count, energy fed on test day, protein fed on test day and dry matter fed on test day etc.). The disadvantage of these models is that they require too many biological parameters requiring a large scale, expensive and time consuming data recording and collection scheme. Unfortunately, this data is unavailable for typical pasture-based dairy farms at the practical level limiting the usability of these ANN models.

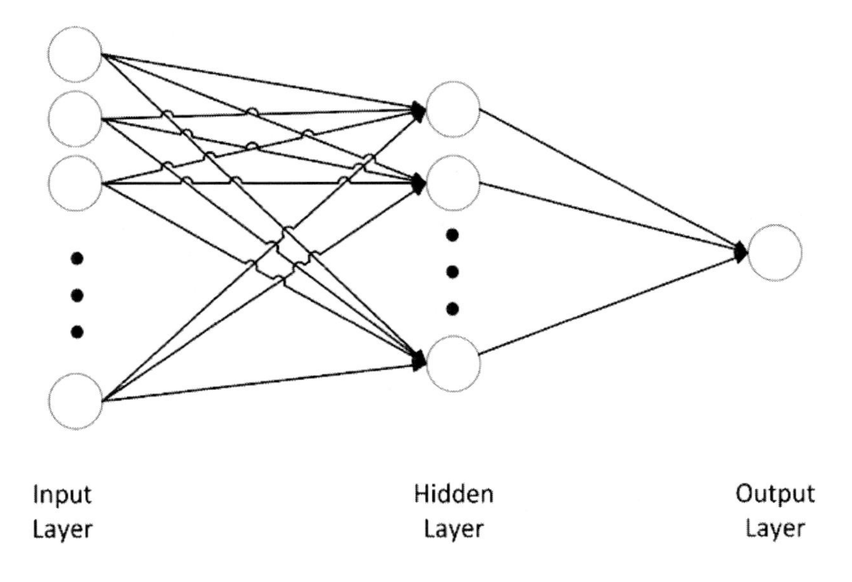

Figure 5. A typical feedforward ANN.

Dynamic ANN Model

Neural networks are generally classified into two types: static (non-recurrent) and dynamic (recurrent) networks (Medsker and Jain, 2001). In contrast to static neural networks, dynamic neural networks (DNN) may result in good time-series prediction performance due to their embedded memory capability (retaining information to be used at later time step) (Connor et al., 1994; Von Zuben and de Andrade Netto, 1995). The internal strategy of a static feedforward (FFD) multilayer ANN is that all outputs are generated from current inputs, however, outputs of a DNN are based on both current and previous inputs and outputs. This short-term memory mechanism may enhance the whole networks' performance on learning and recognition. Figure 6 shows a typical DNN with a feedback loop from the output back to the input layer.

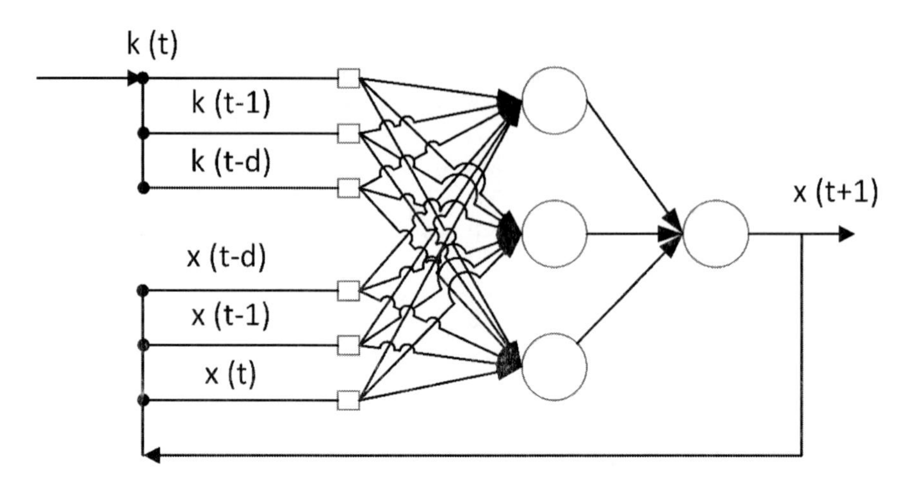

Figure 6. A typical DNN (Murphy et al., 2014).

Tapped delay lines (TDL) are placed before the input and the feedback loop from the output to the hidden layer, and used to delay the input signal by a number of time steps. In many applications to date, dynamic neural networks are referred to as the nonlinear auto-regressive model with exogenous input (NARX) due to the exogenous (external data) feedback element and TDLs. The mathematical representation of a NARX model is as follows:

$$\chi(t+1) = f(\chi(t), \chi(t-1) \ldots, \chi(t-n), \kappa(t), \kappa(t-1) \ldots, \kappa(t-n)) \tag{21}$$

The embedded short term memory within the NARX model allows for an increased efficiency in back propagating gradient information compared to other ANN models. In particular, NARX models have been shown to perform well at recognising short-term patterns in the input data (Lin et al., 1997, 1998).

The NARX model has been proven to be a powerful tool for short term time series analysis in chaotic and noisy environments (Diaconescu, 2008a; Mirzaee, 2009) and time series prediction (Barbounis et al., 2006; El-

Shafie et al., 2012; Khoshnevisan et al., 2014; Paoli et al., 2010; Voyant et al., 2011).

The NARX model could be presented as a more accurate alternative to conventional regression modelling techniques, especially for short-term milk yield predictions (Murphy et al., 2014). The study of Murphy et al. demonstrated that the NARX model was successful in milk production forecasting at the herd level with training data consisting of herd DMY, DIM and the NCM. In this study, the NARX model was compared with a static ANN model and a MLR model using three years of historical milk production data. The result comparison was based on prediction of the total daily herd milk yield over a whole lactation using different forecast horizons. The NARX was found to increase the prediction accuracy as the horizon was shortened from 305 to 50, 30 and 10 days, while the other two models could not reduce error to the same extent due to lack of ability to dynamically learn from their errors from previous predictions. On the other hand, this study claimed that it is difficult to compare the effectiveness of models for individual cows due to lack of specific information, especially for a dairy farm without the use of a sophisticated computerized milk recording system. In addition, this study also indicated that each mentioned study used case-specific data to predict milk yield for a herd or cow at the unique conditions and it was probably difficult to conduct comparisons over different studies.

Mechanistic Approach

Mechanistic approaches offer insights into the mammary gland physiological processes and thus, offer an increase in biological parameters (Grossman and Koops, 2003; Neal and Thornley, 1983; Pollott, 2000). However, according to study of Pollott (2000), one limitation of mechanistic models is that they cannot fit the data well based on current monthly milk records and are often over parameterized. A recent study proposed an explanatory mathematical and biological model based on udder physiology which could be tested and validated using empirical data

(Gasqui and Trommenschlager, 2017). The author claimed that this model could both predict lactation traits (such as the length of peak lactation, peak milk yield, and total milk yield), explore a physiological process and pinpoint potential problems. This model enhanced Wood's model, which does not have a clear biological interpretation.

Recently, a mechanistic model was proposed by Baudracco et al. (2012) to predict milk production of a pasture-based dairy cow. This model comprised of three previously published models including an 'INTAKE'' model that predicts herbage dry matter (DM) intake by grazing dairy cows, a 'MILK' model that predicts potential milk yield and a 'LIPID' model that predicts genetically driven live weight (LW) and body condition score (BCS). The highlight of this model is that it could give prediction results with satisfactory accuracy (concordance correlation coefficient value equalling 0.76 for milk yield), under the conditions that all input parameters for three sub-models should be provided meaning this model requires large amounts of detailed input information.

Another recent mechanistic model was proposed by Ruelle et al. (2015) to predict milk production of pasture-based dairy herd. This model integrates three components, including a herd dynamic milk model that predicts the production of standard milk at 4.0% fat and 3.1% protein, a paddock model that predict grass conditions and grazing management rules that simulate the impact of dairy farm management rules. The advantage of this model is that it is able to take into account the management effect on dairy farms and has the ability to integrate updated grass growth models and bring wider usability. However, with the same innate characteristics as that of Baudracco et al.'s model, the requirement of embracive on farm data may reduce the practicability and limit the application of model.

Standard Lactation Curve Method in Ireland

The Standard Lactation Curve (SLAC) method was proposed by Olori and Galesloot (1999) and is currently used in Ireland for predicting milk yield. This method was developed through interpolating 341,652 lactations from 121,179 cows in 5,225 herds, which has a library of three equations as follows (Equation 22, 23, 24). The SLAC method involved three developmental steps. Step 1) standard lactation curves were derived for each contemporary group of cows defined to suit the production environment (Equation 22). Step 2) 15 milk yield values were predicted at 20 day intervals between the 10th and 290th lactation day using the derived lactation curve (Equation 23) and step 3) the lactation curve was expanded by calculating the milk yields in the unknown sections of the curve based on neighbouring fixed days (Equation 24).

$$Y_n = E(Y_n) + b_1*[Y_{p305} - E(Y_{p305})] + b_2*[Y_k - E(Y_k)] \quad (22)$$

Where Y_n is the predicted yield for day n of the lactation in progress, $E(Y_n)$ is the expected yield on day n from the SLAC, Y_{p305} is the realised 305-day yield of the previous lactation, $E(Y_{p305})$ is the expected 305-day yield of the previous lactation, Y_k is the yield on the last test day k of the lactation in progress, $E(Y_k)$ is the expected yield on the last test day k from the SLAC, and b_1 and b_2 are the lactation projection factors. Projection factors were derived by recurrent regression analyses involving the deviation of the yield on the last test and the previous lactation from their expectations. Additionally, for predicting fixed days before the first test by back prediction, a revised equation is used where Y_k is equal to the yield on the first test day.

The prediction yield for the fixed DIM is calculate by interpolation using following equation:

$$Y_n = G_n + [(Y_2-Y_1) - (G_2 - G_1)] / [(X_2-X_1) * (X_n - X_1)] + (Y_1 - G_1) \quad (23)$$

Where Y_n is the yield to be predicted, Y_2 and Y_1 are the observed daily yields, X_1 and X_2 are the days when Y_1 and Y_2 were measured respectively, X_n is the day for which a yield is to be predicted where $X_1 < X_n < X_2$, and, G_n, G_1, G_2 are the expected yields $E(Y)$ on days n, 1 and 2, respectively.

Once the yields of each of the fixed days have been calculated, the cumulative 305-day milk production (fat or protein) yield can be calculated using Equation 24:

$$Y_{305} = \sum_{i=i}^{n} 0.5\,[Y_i*(int_i - 1) + Y_{i+1}*(int_i + 1)] \tag{24}$$

Where Y_{305} is the 305-day milk production yield, Y_i is the yield of day i, int_i is the interval in days between the daily yields Y_i and Y_{i+1}, n is total number of daily yields (measured and predicted).

As states above, results of the SLAC method were based on a large number of sample data and confirmed that the correlations between projected and actual whole lactation yields increased with progressing length of records. The projection process was able to differentiate cows with potential from those without potential to produce further in projecting short lactations.

Model Application and Comparison

Several studies have been carried out comparing the milk yield prediction performance of the modelling techniques discussed in section 2.2. Similar studies have been carried out comparing the forecast accuracy of each individual model category. In particular, numerous works have been carried out comparing model performance within two categories, across different modelling techniques and results evaluation platforms (Adediran et al., 2012; Bhosale and Singh, 2017; Cole et al., 2009; Druet et al., 2003; Gandhi et al., 2010; Grzesiak et al., 2003; Murphy et al., 2014; Olori et al., 1999; Otwinowska-Mindur et al., 2013; Quinn et al., 2005; Sharma and Kasana, 2006; Silvestre et al., 2006).

Olori et al.(1999) analysed five empirical models for milk yield forecasting of stall fed cows between 1990 and 1994 in the UK. These five standard lactation curve models were developed and analysed for their prediction capabilities of the average daily milk yield of 325 first lactation cows in a single herd. Weekly averages of daily milk yield were obtained from a single Holstein-Friesian herd, and used for developing the standard lactation curve models. Hence, this study focused on analysing model performance with a relatively low data variance. Based on the adjusted R-squared correlation (R^2) and the root mean square error (RMSE), the herd average milk yield was predicted with a high degree of accuracy by all models ($0.99 > R^2 > 0.94$, 0.67 kg> RMSE >0.17 kg). For predicting of individual lactations, the mean and standard deviation of R^2 for individual lactation predicted was 0.66±0.25, 0.69±0.24, 0.65±0.25 and 0.67±0.24 for the incomplete gamma (Wood, 1967), exponential (Wilmink, 1987), inverse polynomial (Nelder,1966; Yadav et al., 1977), and mixed log (Guo and Swalve, 1995) models, respectively. Results showed the models fitted equally well for typical lactations which peeked between the 6th and 9th week ($0.76 > R^2 > 0.70$) and fitted equally poorly for non-typical lactations ($0.69 > R^2 > 0.20$). Thus, the accuracy levels of the model predictions depended upon the variance of the data utilised for fitting the lactation curves as opposed to the specific characteristics of the model. I.e., an increased number of cows following the typical lactation pattern will result in an increased prediction accuracy. In another words, the suitability of models predicting herd lactation curves depend on the function utilised whereas the suitability of a model predicting individual cow lactation curves depend upon the biological nature of the training lactation data which varies randomly between cows. In order to analyse the performance of curve fitting models for milk yield prediction, high resolution data is required because all models performed equally well.

Druet et al. (2003) compared splines with traditional polynomial models (polynomial regression (Ali and Schaeffer, 1987), exponential (Wilmink,1987), Legendre polynomial, regression splines (White et al., 1999)) for modelling the fixed part of the lactation curve as well as the genetic parameters of Holsteins cows using 1.69 million first lactation records between 1994 and 2000 in France. The evaluation of each model was based upon the model fitting accuracy and flexibility and computational difficulty. Different performance rankings were obtained according to two criterions used in this study: fixed classes curves performed better than the others based on the mean sum of squares of the residuals (MSSE = 1187 kg), while the regression spline were the best based on the mean residual (<1 in each 305 DIM). As the authors mentioned, the size of a milk yield dataset may explain differences between the results of this study (0.8 million records in the first lactation) and contemporaneous studies. Thus, it is difficult to compare the accuracy of these lactation curve models to those developed utilising a smaller number of cows, due to the scale of this study. In particular, this is relevant from an Irish perspective where the overall number of dairy cows equalled 1.3 million by the end of December 2016 (Central Statistics Office, 2017). Even though a recent model comparison study utilising 4.5 million milk records of Polish Holstein-Friesian cows from 530,425 lactations (Otwinowska-Mindur et al., 2013) may have been compared with the study of Druet et al. (2003). Five models were compared in the study in Poland, including exponential (Wilmink, 1987), polynomial regression (Ali and Schaeffer, 1987), mixed log (Guo and Swalve, 1995), third-order Legendre polynomials, and fourth-order Legendre polynomials. According to two criterions used in Druet et al.'s study (the mean absolute error (MAE) and the mean square error (MSE)), the polynomial regression model performed best for either the 305-day lactations (1.32 kg - 1.55 kg in MAE and 3.52 kg - 4.93 kg in MSE) or extended 400-day lactations (1.39 kg - 1.62 kg in MAE and 4.33 kg - 5.63 kg in MSE). Both studies failed to compare similar models, thus, results from each study are highly unique and case-specific, different studies analysing different number of milking records impedes the ability to compare research results. Beyond objective limitation, model selection in each study was subjective and attended by more or less preference of the author, hence not all studies utilise the same models. This is the second constraint of comparison of outputs from different studies.

Similarly, Quinn et al. (2005) compared 14 empirical algebraic models for Irish pasture-based milk yield data (14,965 records) between 1995 and 2001. Tested models consist of the most commonly used models in previous studies, i.e., including exponential (Wilmink, 1987), polynomial regression (Ali and Schaeffer, 1987), mixed log (Guo and Swalve, 1995), incomplete gamma (Wood, 1967), exponential (Wilmink, 1987), inverse polynomial (Yadav et al., 1977) and etc. The mean square prediction error (MSPE) and R^2 value were used to compare the model performance. The polynomial regression model was found to be the best on the basis of its MSPE (501.7) and R^2 (0.68) with the 5,937 kg estimated annual yield and percentage deviation (3.9%), in contrast, the Ali-B model which was proposed in this study, has a relative level of MSPE (520.9) and R^2 (0.67), but has the smallest difference between the 5,795 kg estimated annual yield and the actual average annual yield (5,702 kg) and percentage deviation (1.6%). Quinn et al. found that using a curve fitting model to predict the milk yield for an individual cow always required parameter adjustments due to many regional effects such as climate, soil quality and environment. Beyond model comparison, differences of parameter estimates for the same model resulting from differences between datasets from experimental herds and commercial herds have been considered. The author demonstrated that there were two significant differences in prediction results based on two training datasets obtained from experimental herds in 1978 and commercial herds in 2003: 1) average annual yield per cow has increased from 2,364 kg to 5,448 kg and 2) the week where peak yield occurred shifted from week six to week eight.

Silvestre et al. (2006) reviewed seven models including polynomials, Legendre polynomials and cubic splines models using data from stall based dairy cows collected between 1999-2001 in Portugal. The dataset consisted of 144 complete lactations (305-days) of 139 cows and the criteria consisted of mean error, standard deviation (SD) of error correlation, the quotient (Q) between the error sum of squares and the observed sum of squares, and etc. The cubic splines model showed better prediction performance (mean error <1, SD of error < 5, $R > 0.83$ and mean of $Q < 4.1$, SD of $Q < 4.0$), when compared to the including exponential (Wilmink, 1987), polynomial regression (Ali and Schaeffer, 1987), incomplete gamma (Wood, 1967) models. All seven models achieved better prediction using shorter interval data from calving to first test day (less than 30-day vs more than 60 day), and these results showed that the differences in prediction accuracy between models became more significant as the amount of data decreased and the timing of the initiation of data collection was delayed. In particular, the polynomial models were highly affected by the reduction the sample dimension. This study agrees that the performance of polynomial models depends on both the sampling properties and the variability between each individual cow within test samples. Also, the limitation of the data collection frequency from an economical view was mentioned, however, the impact of this limitation may be reduced as more advanced automatic milking systems are deployed in modern dairy farms (O'Brien et al., 2015).

For forecasting milk yields during long lactation cycles (>= 500 days) Cole et al. (2009) utilised 152,734 cows as sample consisting of six breeds. After editing, 348,123 lactation records were used for parameter estimation and random samples of one million records from US Holsteins were abstracted for validation purposes. In this study, the 7-day average milk yield was used as the daily milk data. Average milk yield and SD at any DIM were estimated by utilising an incomplete gamma (Wood, 1967) and compared with actual observation yields using correlations. As a result, cows with long lactations had different shapes compared to those of 305-day lactations and the author stressed that using only 305-day lactation records may produce opposite results and data used in this study should be as least 500-day. In contrast, the same data were used in another study of modelling long lactations based on the comparison of 305-day and 999-day lactations (Dematawewa et al., 2007). This study compared nine models including incomplete gamma (Wood, 1967), exponential (Wilmink, 1987) based on several criteria such as: error of squares (SSE), square root of mean square error (RMSE), adjusted squared correlation (R^2). The results showed that the prediction of incomplete gamma model for 999-day were the best with respect to RMES and R^2 (7.175 kg, 0.076 for 305-day lactations of first parity cows, 9.380 kg, 0.270 for 305-day lactations of cows in other parities, 7.811 kg, 0.210 for 999-day lactations of first parity cows, 9.623 kg, 0.367 for 999-day lactations of cows in other parities). However, the point is that the possible comparison can be conducted based on the results from jointing these two studies which have common test data and that's the most valuable contribution for any follow-up study.

Grzesiak et al. (2003) presented a comparision between the static ANN model and the MLR model for 305-day lactation yield predicions using data from 902 Polish Holstein-Friesian cows during 1994 to 1999. For the purpose of training the ANN model and the MLR model, each cow was described with a group of seven input variables, including average cumulative lactation milk yield, DIM, average milk yield of first four months, and numerical month of calving (1-12). Model evaluation based on criteria included RMSE, SD, relative mean error of prediction (MEP) and R^2. By training these data, both the MLR model and static ANN model obtained good prediction performance with an R^2 value of 0.87 for the MLR and 0.88 for the ANN model. The SD ranged between 0.36 and 0.39 for the MLR model, while the SD ranged between 0.34 and 0.35 for the ANN model. These results implied the ANN model can be an alternative to the conventional MLR model. Another study carried out by Grzesiak et al. (2006) compared the static ANN model with Wood's gamma model (Wood, 1967) using datasets consisting of 137,507 daily records of 320 cows over 2000-2002. In this study, the ANN model was trained with a group of five variables: the HF percentage, the age at calving in months, the numerical month of calving (1-12), DIM, and the lactation number (1-3), while the Wood's model was only trained with DIM and average daily milk yield. Based on the same criteria as in their previous study (RMSE, R^2), the R^2 value of the static ANN model was 0.77, in contrast to the values of the Wood's model which ranged from 0.45 to 0.62. The forecasting improvement of the static ANN model was contributed to the ability to use additional variable inputs derived from the population of test cows. In this study, to take the same dimension of input data of the static ANN model, the pre-processing of training data for the Wood's model was complicated. This pre-processing divided the raw dataset into different groups such as age groups, genetic groups, calving season groups, lactation groups and etc. resulting in the production of 24 equations. The authors implied that it was virtually impossible to repeat this for a single farm and the ANN model was the optimum solution for this kind of study. However there was no further test to combine these two studies, thus, the conclusions are only valid for each study respectively.

Sharma et al. (2006) proposed and compared two static ANN models with a conventional MLR model based on the prediciton of the first lacation 305-day milk yield using filtered data from raw records of 672 Indian Karan Fries dairy cows. In this study, the training data was collected over a period of 20 years (1982-2002) and adjusted values of input included weight at maturity,age at calving, peak milk yield and days to attain the peak milk yield. Although there may be a variation in the performance of cows due to the effect of various non-genetic factors, the variation may not be significant enough to be detected due to the small amount of sample cows distributed over 20 years. Based on percentage RMSE value, the results of this study showed that one static ANN (radial basis function neural networks, RMSE = 9.44%) performs relatively better than MLR model (RMSE = 9.46%). Similarly, the other static ANN (back propagation neural networks, RMSE = 11.22%) performs more or less equivalently. Subsequent studies found that the ANN was more accurate than MLR model on prediction of life time milk yield on the basis of the first lactation traits using data of Sahiwal cattle (Bhosale and Singh, 2017) and Holstein-Friesian dairy cows (Gandhi et al., 2010), respectively. Although studies of Bhosale and Singh and the study of Gandhi et al. uesd RMSE and R^2 as comparison criterion, these studies compared ANN models and MLR models using different empirical data of multiple breeds utilising data over various periods, hence the conclusions are qualitatively consistent, however with quantitative differences in detail.

Adediran et al. (2012) analysed 16 models including empirical models and semiparametric models using data from pasture-based dairy cows collected between 1998-2007 in the Australian states of Tasmania and Victoria (96,747 records from 11,643 lactations). Both average and individual cow lactations were used for model evaluation. Based on these datasets and evaluation criteria including residual mean square (RMS), SD of RMS, mean error, SD of mean error and R^2, models with biologically interpretable parameters were found to have good performance, compared to the polynomial model and the gamma model. I.e., the log-quadratic model (Adediran et al., 2012) showed a high R^2 value (0.99) of prediction for individual cow lactations, with a low R^2 (0.18) value for the incomplete gamma (Wood, 1967), and in addition, the RMS values for these two models were 0.03 and 16.9, respectively. This study confirmed the effect of the day at the first test day and number of recorded test days on the fitting performance of lactation models. The overall goodness of fit of all lactation models were adequate while the most accurate model was the log-quadratic model which was recommended for fitting test day milk yield. However, the limitation of this study is that model accuracy was tested on data from another dairy system (stall-based farms), thus, the robustness of the log-quadratic model was not quantified for pasture-based systems.

In a recent study by Murphy et al. (2014), the MLR model, the static ANN model and the NARX model were compared using Irish pasture-based data collected from 140 Holstein-Friesian cows between 2006 and 2010. The NARX model was introduced as an advanced ANN model compared to the conventional static ANN model. Inconsistent with the previous study (Grzesiak et al., 2003) and using the same evaluation criteria including RMSE, and R^2, over the full 305-day cycle with four different horizons ranging from 305-day to 10-day, the static ANN did not produce superior prediction in compared to the MLR model: the RMSE of the static ANN forecast decreased from 12.03% to 10.7% and R^2 increased from 0.889 to 0.911, while the RMSE of the MLR forecast ranged from 10.62% to 10.54% and R^2 decreased from 0.917 to 0.916. In contrast, the NARX proved to have considerably better accuracy for predicting milk yield for different horizons. In particular, the prediction error dropped monotonically in correspondence with the shortening of the prediction horizon, the RMSE of the NARX forecast decreased from 8.59% to 5.84% and R^2 increased from 0.936 to 0.968. In this study, the forecast accuracy of the static ANN was not better than the MLR model and this may be caused by data limitation as only DIM, NCM and DHMY were selected as the training inputs. In spite of this, the NARX still produced best prediction accuracy and the error reduced in accordance with the shortening of the prediction horizon. This attribute was due to the NARX model's ability to adapt and update its trajectory based on past errors. The authors stated that it is difficult to compare the results of this study with previous studies due to each study using case-specific data, each uniquely impacted by environmental, grazing and feeding factors.

From the discussion above, results from these studies show varying levels of accuracy for different models with different data inputs, for specific applications. For example, the Ali and Schaeffer model (Ali and Schaeffer, 1987) was the most accurate model for Canadian datasets in the original study, while the revised Ali-B model (Quinn et al., 2005) was found to have better performance than the original Ali and Schaeffer model. From an Irish perspective, the Ali and Schaeffer model was found better than the Ali-B model again (Zhang et al., 2014), while the log-quadratic model (Adediran et al., 2012) did not perform as well on Irish data as it did on Australian data. Concurrently, there may be a potential opportunity to discover a model which has better prediction performance than the NARX model at specific prediction horizons for Irish data. In addition, in comparing these configurations within the same model category or within cross-category may increase the complexity and time consumption of the overall development of the prediction model. Furthermore, cross category milk yield model comparisons are technically and computationally more complex than those within the same category, as discussed above. Although many studies exist comparing the prediction performance of these models using average data of individual cows (Cole et al., 2009; Madouasse et al., 2010; Van Bebber et al., 1999), a comparison focusing on an individual cow level has yet to be carried out. Milk yield forecasting at an individual cow level could be beneficial to numerous applications in dairy industry including monitoring health conditions and disease detection by monitoring individual cow milk yield, i.e., udder mastitis (Andersen et al., 2011; Gasqui and Trommenschlager, 2017); decision support for advanced milking parlours and milking machines (Thomas and DeLorenzo, 1994) and precision input for herd simulation models (Petek and Dikmen, 2006).

Conclusion

A review related to milk yield prediction was presented and investigated within this chapter, covering different lactation modelling approaches and model application and comparison. Numerous modelling techniques have been proposed and applied to milk yield forecasting, including classical curve fitting models, regressive models, auto-regressive models and mechanistic models. Due to the main limitation that specific milk yield datasets are highly case specific, most models could be the optimal model based on the specific research objects and test datasets under unique conditions. Therefore, researchers from similar or dissimilar regions do not have a mutual target to compare with. Most models are developed and adapted for their countries under numerous limitations. As a result, it is difficult to propose a standard or the optimal model for dairy cows for whatever vertical comparison (different regions over the same period) or horizontal comparisons (same region over different periods). There has been consistent growth in the average annual milk yield over the past decades, along with the variations of milk yield in the lactations of individual breeds. When including advanced feeding methods, new grazing management techniques, climate changes and and changes in genetic level, it is reasonable to assume continued changes in mean milk yield and variance level. Therefore, the future efficacy of classical milk forecasting techniques is uncertain and may lead to the mass adoption of more modern adaptive techniques.

References

Adediran, S. A. et al., 2012. Comparative evaluation of a new lactation curve model for pasture-based Holstein-Friesian dairy cows. *Journal of Dairy Science*, 95(9), pp. 5344–5356. Available at: http://linkinghub.elsevier.com/retrieve/pii/S0022030212005516.

Ali, T. E. & Schaeffer, L. R., 1987. Accounting for covariances among test day milk yields in dairy cows. *Canadian Journal of Animal Science*, 67(3), pp.637–644.

Andersen, F. et al., 2011. Mastitis and the shape of the lactation curve in Norwegian dairy cows. *Journal of Dairy Research*, 78(1), pp. 23–31. Available at: http://www.ncbi.nlm.nih.gov/pubmed/21118610 [Accessed July 26, 2017].

André, G. et al., 2010. Increasing the revenues from automatic milking by using individual variation in milking characteristics. *Journal of Dairy Science*, 93(3), pp.942–953. Available at: http://linkinghub.elsevier.com/retrieve/pii/S0022030210000615 [Accessed July 28, 2017].

Barbounis, T. G. et al., 2006. Long-Term Wind Speed and Power Forecasting Using Local Recurrent Neural Network Models. *IEEE Transactions on Energy Conversion*, 21(1), pp.273–284. Available at: http://ieeexplore.ieee.org/xpl/articleDetails.jsp?tp=&arnumber=1597347&queryText=Long-term+wind+speed+and+power+forecasting+using+local+recurrent+neural+network+models.

Baudracco, J. et al., 2012. e-Cow: an animal model that predicts herbage intake, milk yield and live weight change in dairy cows grazing temperate pastures, with and without supplementary feeding. *Animal*, 6(6), pp.980–993. Available at: http://www.journals.cambridge.org/abstract_S1751731111002370.

Van Bebber, J. et al., 1999. Monitoring daily milk yields with a recursive test day repeatability model (Kalman filter). *Journal of dairy science*, 82(11), pp.2421–2429.

Bhosale, M. D. & Singh, T. P., 2017. Development of Lifetime Milk Yield Equation Using Artificial Neural Network in Holstein Friesian Cross Breddairy Cattle and Comparison with Multiple Linear Regression Model. *Current Science*, 113(5), pp.951–955. Available at: http://www.i-scholar.in/index.php/CURS/article/view/158480 [Accessed October 20, 2017].

Breen, M., Murphy, M. D. & Upton, J., 2015. Development and validation of photovoltaic and wind turbine models to assess the impacts of renewable generation on dairy farm electricity consumption. In *2015*

ASABE International Meeting. American Society of Agricultural and Biological Engineers, p. 1. Available at: https://elibrary.asabe.org/azdez.asp?JID=5&AID=46275&CID=norl2015&v=&i=&T=1 [Accessed May 6, 2016].

Brody, S., Ragsdale, A. C. & Turner, C. W., 1923. the Rate of Decline of Milk Secretion With the Advance of the Period of Lactation. *The Journal of general physiology*, 5(4), pp.441–4. Available at: http://www.pubmedcentral.nih.gov/articlerender.fcgi?artid=2140574&tool=pmcentrez&rendertype=abstract.

Brody, S., Turner, C. W. & Ragsdale, A. C., 1924. The Relation between the Initial Rise and the Subsequent Decline of Milk Secretion Following Parturition. *The Journal of general physiology*, 6(5), pp.541–5. Available at: http://www.pubmedcentral.nih.gov/articlerender.fcgi?artid=2140670&tool=pmcentrez&rendertype=abstract [Accessed October 24, 2017].

Brotherstone, S., White, I. M. S. & Meyer, K., 2000. Genetic Modelling of Daily Milk Yield Using Orthogonal Polynomials and Parametric Curves. *Animal Science*, 70, pp.407–416. Available at: http://www.researchgate.net/publication/235797516_Genetic_Modelling_of_Daily_Milk_Yield_Using_Orthogonal_Polynomials_and_Parametric_Curves [Accessed October 27, 2015].

Central Statistics Office, 2017. *StatBank.* Available at: http://www.cso.ie/px/pxeirestat/statire/SelectVarVal/Define.asp?Maintable=AKM01&PLanguage=0 [Accessed August 6, 2017].

Cole, J. B., Null, D. J. & VanRaden, P. M., 2009. Best prediction of yields for long lactations. *Journal of Dairy Science*, 92(4), pp.1796–1810. Available at: http://dx.doi.org/10.3168/jds.2007-0976 [Accessed October 2, 2014].

Connor, J. T., Martin, R. D. & Atlas, L. E., 1994. Recurrent Neural Networks and Robust Time Series Prediction. *IEEE Transactions on Neural Networks*, 5(2), pp.240–254. Available at: http://ieeexplore.ieee.org/document/279188/ [Accessed October 18, 2017].

Dematawewa, C. M. B., Pearson, R. E. & VanRaden, P. M., 2007. Modeling Extended Lactations of Holsteins. *Journal of Dairy Science*,

90(8), pp.3924–3936. Available at: http://www.sciencedirect.com/science/article/pii/S0022030207718490 [Accessed October 20, 2014].

Department of Agriculture Food & the Marine, 2016. *Food Wise 2025*. Available at: https://www.agriculture.gov.ie/foodwise2025/ [Accessed August 7, 2017].

Diaconescu, E., 2008. Prediction of Chaotic Time Series with NARX Recurrent Dynamic Neural Networks. In *ICAI'08*. Stevens Point, Wisconsin, USA: World Scientific and Engineering Academy and Society (WSEAS), pp. 248–253. Available at: http://dl.acm.org/citation.cfm?id=1411620.1411669 [Accessed June 3, 2014].

Dongre, V. B. et al., 2012. Comparative efficiency of artificial neural networks and multiple linear regression analysis for prediction of first lactation 305-day milk yield in Sahiwal cattle. *Livestock Science*, 147(1–3), pp.192–197. Available at: http://www.researchgate.net/publication/257700789_Comparative_efficiency_of_artificial_neural_networks_and_multiple_linear_regression_analysis_for_prediction_of_first_lactation_305-day_milk_yield_in_Sahiwal_cattle [Accessed October 29, 2015].

Druet, T. et al., 2003. Modeling Lactation Curves and Estimation of Genetic Parameters for First Lactation Test-Day Records of French Holstein Cows. *Journal of Dairy Science*, 86(7), pp.2480–2490. Available at: http://linkinghub.elsevier.com/retrieve/pii/S0022030203738429 [Accessed July 26, 2017].

El-Shafie, A. et al., 2012. Dynamic versus static neural network model for rainfall forecasting at Klang River Basin, Malaysia. *Hydrol. Earth Syst. Sci.*, 16(4), pp.1151–1169. Available at: http://www.hydrol-earth-syst-sci.net/16/1151/2012/ [Accessed June 3, 2014].

Esen, H. et al., 2008. Performance prediction of a ground-coupled heat pump system using artificial neural networks. *Expert Systems with Applications*, 35(4), pp.1940–1948. Available at: http://www.sciencedirect.com/science/article/pii/S0957417407004046 [Accessed June 1, 2014].

Fukushima, K., 1988. Neocognitron: A hierarchical neural network capable of visual pattern recognition. *Neural Networks*, 1(2), pp.119–130. Available at: http://www.sciencedirect.com/science/article/pii/0893608088900147 [Accessed April 24, 2017].

Gandhi, R. S. et al., 2010. Artificial Neural Network versus Multiple Regression Analysis for Prediction of Lifetime Milk Production in Sahiwal Cattle. *Journal of Applied Animal Research*, 38(2), pp.233–237. Available at: http://www.tandfonline.com/doi/abs/10.1080/09712119.2010.10539517 [Accessed October 23, 2017].

Gasqui, P. & Trommenschlager, J. M., 2017. A new standard model for milk yield in dairy cows based on udder physiology at the milking-session level. *Scientific reports*, 7(1), p.8897. Available at: http://www.ncbi.nlm.nih.gov/pubmed/28827751 [Accessed September 26, 2017].

Gorgulu, O., 2012. Prediction of 305-day milk yield in Brown Swiss cattle using artificial neural networks. *South African Journal of Animal Science*, 42(3).

Green, P. J. & Silverman, B. W., 1993. *Nonparametric Regression and Generalized Linear Models: A roughness penalty approach*, CRC Press. Available at: http://books.google.ie/books?id=-AIVXozvpLUC.

Green, P. J. & Silverman, B. W., 1994. Nonparametric Regression and Generalized Linear Models: A Roughness Penalty Approach. *Monographs on statistics and applied probability*, 58(3), p.182. Available at: http://www.jstor.org/stable/1269920?origin=crossref.

Grossman, M. & Koops, W. J., 2003. Modeling extended lactation curves of dairy cattle: a biological basis for the multiphasic approach. *Journal of dairy science*, 86(3), pp.988–98. Available at: http://linkinghub.elsevier.com/retrieve/pii/S0022030203736820 [Accessed June 26, 2017].

Grzesiak, W. et al., 2003. A comparison of neural network and multiple regression predictions for 305-day lactation yield using partial lactation records. *Canadian Journal of Animal Science*, 83(2), pp.307–310.

Grzesiak, W., Błaszczyk, P. & Lacroix, R., 2006. Methods of predicting milk yield in dairy cows-Predictive capabilities of Wood's lactation curve and artificial neural networks (ANNs). *Computers and Electronics in Agriculture*, 54(2), pp.69–83. Available at: http://www.sciencedirect.com/science/article/pii/S0168169906000998 [Accessed October 20, 2014].

Guo, Z. & Swalve, H. H., 1995. Modelling of the lacation curve as a sub-model in the evaluation pf test day record. *Interbull meeting*, (36), pp.52–57. Available at: http://agris.fao.org/agris-search/search.do?f=1996/SE/SE96003.xml;SE9610776.

Hocaoĝlu, F. O., Gerek, Ö. N. & Kurban, M., 2007. A Novel 2-D Model Approach for the Prediction of Hourly Solar Radiation. In F. Sandoval et al., eds. *Lecture Notes in Computer Science.* Springer Berlin Heidelberg, pp. 749–756. Available at: http://link.springer.com/chapter/10.1007/978-3-540-73007-1_90 [Accessed June 1, 2014].

Ince, D. & Sofu, A., 2013. Estimation of lactation milk yield of Awassi sheep with Artificial Neural Network modeling. *Small Ruminant Research*, 113(1), pp.15–19. Available at: http://dx.doi.org/10.1016/j.smallrumres.2013.01.013.

Jones, T., 1997. Empirical Bayes Prediction of 305-Day Milk Production. *Journal of Dairy Science*, 80(6), pp.1060–1075. Available at: http://www.sciencedirect.com/science/article/pii/S0022030297760314 [Accessed September 30, 2015].

Jung, S. & Kim, S. S., 2007. Hardware Implementation of a Real-Time Neural Network Controller With a DSP and an FPGA for Nonlinear Systems. *IEEE Transactions on Industrial Electronics*, 54(1), pp.265–271. Available at: http://ieeexplore.ieee.org/xpl/login.jsp?tp=&arnumber=4084734&url=http%3A%2F%2Fieeexplore.ieee.org%2Fxpls%2Fabs_all.jsp%3Farnumber%3D4084734.

Kalogirou, S. A. & Bojic, M., 2000. Artificial neural networks for the prediction of the energy consumption of a passive solar building. *Energy*, 25(5), pp.479–491. Available at: http://www.sciencedirect.com/science/article/pii/S0360544299000869 [Accessed April 24, 2017].

Kerr, D. V. et al., 1998. A study of the effect of inputs on level of production of dairy farms in Queensland - a comparative analysis of survey data. *Australian Journal of Experimental Agriculture*, 38(5), p.419. Available at: http://www.publish.csiro.au/view/journals/dsp_journal_fulltext.cfm?nid=72&f=EA97153 [Accessed September 30, 2015].

Khazaei, J. & Nikosiar, M., 2005. *Approximating milk yield and milk fat and protein concentration of cows through the use of mathematical and artificial neural networks models*. pp.91–105.

Khoshnevisan, B. et al., 2013. *Developing an Artificial Neural Networks Model for Predicting Output Energy and GHG Emission of Strawberry Production*. 3(4), pp.43–54.

Khoshnevisan, B. et al., 2014. Prediction of potato yield based on energy inputs using multi-layer adaptive neuro-fuzzy inference system. *Measurement: Journal of the International Measurement Confederation*, 47(1), pp.521–530. Available at: http://www.sciencedirect.com/science/article/pii/S0263224113004570.

Kim, T. Y. et al., 2004. Artificial neural networks for non-stationary time series. *Neurocomputing*, 61(1–4), pp.439–447.

Kim, Y. H. & Lewis, F. L., 2000. Optimal design of CMAC neural-network controller for robot manipulators. *IEEE Transactions on Systems, Man and Cybernetics Part C: Applications and Reviews*, 30(1), pp.22–31. Available at: http://ieeexplore.ieee.org/document/827451/ [Accessed April 24, 2017].

Kirkpatrick, M., Hill, W. G. & Thompson, R., 1994. Estimating the covariance structure of traits during growth and ageing, illustrated with lactation in dairy cattle. *Genetical Research*, 64(1), p.57. Available at: http://www.journals.cambridge.org/abstract_S0016672300032559.

Kominakis, A. P. P. et al., 2002. A preliminary study of the application of artificial neural networks to prediction of milk yield in dairy sheep. *Computers and Electronics in Agriculture*, 35(1), pp.35–48. Available at: http://www.sciencedirect.com/science/article/pii/S0168169902000510 [Accessed October 20, 2014].

Lacroix, R. et al., 1995. Prediction of Cow Performance with a Connectionist Model. *Transactions of the ASAE*, 38(5), pp.1573–1579. Available at: http://elibrary.asabe.org/abstract.asp??JID=3&AID=27984&CID=t1995&v=38&i=5&T=1 [Accessed January 8, 2017].

Lin, T., Giles, C. L. & Horne, B. G., 1997. Algorithm for NARX Neural Networks. *IEEE Transactions on Signal Processing*, 45(11), pp.2719–2730.

Lin, T. L. T. et al., 1998. What to remember: how memory order affects the performance of NARXneural networks. *1998 IEEE International Joint Conference on Neural Networks Proceedings. IEEE World Congress on Computational Intelligence (Cat. No.98CH36227)*, 2, pp.1051–1056.

Lyons, W. B. et al., 2004. A novel multipoint luminescent coated ultra violet fibre sensor utilising artificial neural network pattern recognition techniques. *Sensors and Actuators A: Physical*, 115(2–3), pp.267–272. Available at: http://www.sciencedirect.com/science/article/pii/S0924424704002365 [Accessed June 1, 2014].

Madouasse, A. et al., 2010. Use of individual cow milk recording data at the start of lactation to predict the calving to conception interval. *Journal of Dairy Science*, 93(10), pp.4677–4690. Available at: http://linkinghub.elsevier.com/retrieve/pii/S0022030210004996 [Accessed August 20, 2016].

Medsker, L. R. & Jain, L. C., 2001. *Recurrent Neural Networks: Design and Applications*, CRC Press. Available at: https://books.google.ie/books?hl=en&lr=&id=ME1SAkN0PyMC&oi=fnd&pg=PA1&dq=recurrent+Neural+Networks:+Design+and+Applications&ots=7bsydM2TWn&sig=M6YSUnNoDVexc6zxv6t7UwT-qlc&redir_esc=y#v=onepage&q=recurrent Neural Networks%3A Design and Applications&f=false [Accessed October 18, 2017].

Melzer, N., Trißl, S. & Nürnberg, G., 2017. Short communication: Estimating lactation curves for highly inhomogeneous milk yield data of an F 2 population (Charolais × German Holstein). *Journal of Dairy Science*, 100(11), pp.9136–9142.

Mirzaee, H., 2009. Long-term prediction of chaotic time series with multi-step prediction horizons by a neural network with Levenberg–Marquardt learning algorithm. *Chaos, Solitons & Fractals*, 41(4), pp.1975–1979. Available at: http://www.sciencedirect.com/science/article/pii/S0960077908003585 [Accessed June 3, 2014].

Motulsky, H. J. & Ransnas, L. a., 1987. Fitting curves to data using nonlinear regression: a practical and nonmathematical review. *FASEB journal : official publication of the Federation of American Societies for Experimental Biology*, 1(5), pp.365–74. Available at: http://www.ncbi.nlm.nih.gov/pubmed/3315805.

Murphy, M. D. et al., 2014. Comparison of modelling techniques for milk-production forecasting. *Journal of dairy science*, 97(6), pp.3352–63. Available at: http://www.sciencedirect.com/science/article/pii/S0022030214002690.

Murphy, M. D., O'Mahony, M. J. & Upton, J., 2015. Analysis of an Optimized Milk Cooling Controller for Dynamic Electricity Pricing Tariffs. In *2015 ASABE International Meeting*. American Society of Agricultural and Biological Engineers, p. 1. Available at: https://elibrary.asabe.org/azdez.asp?JID=5&AID=46192&CID=norl2015&v=&i=&T=1 [Accessed May 6, 2016].

Neal, H. D. S. C. & Thornley, J. H. M., 1983. The lactation curve in cattle: a mathematical model of the mammary gland. *The Journal of Agricultural Science*, 101(2), pp.389–400. Available at: http://journals.cambridge.org/action/displayAbstract?fromPage=online&aid=4597272.

Nielsen, P. P. et al., 2010. Technical note: Variation in daily milk yield calculations for dairy cows milked in an automatic milking system. *Journal of Dairy Science*, 93(3), pp.1069–1073. Available at: http://linkinghub.elsevier.com/retrieve/pii/S0022030210000755 [Accessed July 28, 2017].

O'Brien, B., Foley, C. & Shortall, J., 2015. Robotic milking in pasture-based systems. *Research: Summer 2015*, 10, pp.18–19. Available at: https://www.teagasc.ie/publications/2015/tresearch-summer-2015.php [Accessed October 23, 2017].

Olori, V. E. et al., 1999. Estimating variance components for test day milk records by restricted maximum likelihood with a random regression animal model. *Livestock Production Science*, 61(1), pp.53–63. Available at: http://linkinghub.elsevier.com/retrieve/pii/S0301622699000524 [Accessed September 9, 2016].

Olori, V. E. et al., 1999. Fit of standard models of the lactation curve to weekly records of milk production of cows in a single herd. *Livestock Production Science*, 58(1), pp.55–63. Available at: http://www.sciencedirect.com/science/article/pii/S0301622698001948.

Olori, V. E. & Galesloot, J. B., 1999. Projection of partial location records and calculation of 305-day yields for dairy cattle in the Republic of Ireland. *Proceedings of Interbull Meet. August 26-27, 1999. Zurich, Swizerland. Interbull bulletin*, 22, pp.1–6.

Otwinowska-Mindur, A. et al., 2013. Modeling lactation curves of Polish Holstein-Friesian cows. Part I: The accuracy of five lactation curve models. *Journal of Animal and Feed Sciences*, 22(1), pp.19–25. Available at: http://www.journalssystem.com/jafs/Modeling-lactation-curves-of-Polish-Holstein-Friesian-cows-Part-I-The-accuracy-of-five-lactation-curve-models,66012,0,2.html [Accessed February 6, 2017].

Pahlavan, R., Omid, M. & Akram, A., 2012. Energy input-output analysis and application of artificial neural networks for predicting greenhouse basil production. *Energy*, 37(1), pp.171–176. Available at: http://dx.doi.org/10.1016/j.energy.2011.11.055.

Paoli, C. et al., 2010. *Use of exogenous data to improve an Artificial Neural Networks dedicated to daily global radiation forecasting*. In pp. 49–52. Available at: http://ieeexplore.ieee.org/xpl/login.jsp?tp=&arnumber=5490018&url=http%3A%2F%2Fieeexplore.ieee.org%2Fiel5%2F5482544%2F5489912%2F05490018.pdf%3Farnumber%3D5490018.

Petek, M. & Dikmen, S., 2006. The effects of prestorage incubation and length of storage of broiler breeder eggs on hatchability and subsequent growth performance of progeny. *Czech Journal of Animal Science*, 51(2), pp.73–77.

Pollott, G. E., 2000. A Biological Approach to Lactation Curve Analysis for Milk Yield. *Journal of Dairy Science*, 83(11), pp.2448–2458. Available at: http://www.sciencedirect.com/science/article/pii/S0022030200751368.

Quinn, N., 2005. *Modelling lactation and liveweight curves in Irish dairy cows*. Dublin City University. School of Computing. Available at: http://doras.dcu.ie/18167/ [Accessed April 24, 2017].

Quinn, N., Killen, L. & Buckley, F., 2005. Empirical algebraic modelling of lactation curves using Irish data. *Irish journal of agricultural and food research*, 44(1), pp.1–13.

Rémond, B. et al., 1997. An attempt to omit the dry period over three consecutive lactations in dairy cows. *Annales de Zootechnie*, 46(5), pp.399–408. Available at: http://www.edpsciences.org/10.1051/animres:19970502 [Accessed August 17, 2016].

Ruelle, E. et al., 2016. Development and evaluation of the herd dynamic milk model with focus on the individual cow component. *Animal*, pp.1–12. Available at: http://www.journals.cambridge.org/abstract_S1751731116001026 [Accessed September 19, 2016].

Ruelle, E. et al., 2015. Development and evaluation of the pasture-based herd dynamic milk (PBHDM) model for dairy systems. *European Journal of Agronomy*, 71, pp.106–114. Available at: http://www.sciencedirect.com/science/article/pii/S1161030115300241 [Accessed September 24, 2015].

Salehi, F., Lacroix, R. & Wade, K. M., 1998. Improving dairy yield predictions through combined record classifiers and specialized artificial neural networks. *Computers and Electronics in Agriculture*, 20(3), pp.199–213. Available at: http://www.sciencedirect.com/science/article/pii/S0168169998000180 [Accessed October 20, 2014].

Sanzogni, L. & Kerr, D., 2001. Milk production estimates using feed forward artificial neural networks. *Computers and Electronics in Agriculture*, 32(1), pp.21–30. Available at: http://www.sciencedirect.com/science/article/pii/S016816990100151X.

Schaeffer, L. R., 2004. Application of random regression models in animal breeding. *Livestock Production Science*, 86(1–3), pp.35–45. Available at: http://www.sciencedirect.com/science/article/pii/S030162260300 1519 [Accessed January 12, 2014].

Shalloo, L. et al., 2004. Description and validation of the Moorepark Dairy System Model. *Journal of dairy science*, 87(6), pp.1945–1959. Available at: http://dx.doi.org/10.3168/jds.S0022-0302(04)73353-6 [Accessed September 19, 2016].

Shalloo, L., Creighton, P. & O'Donovan, M., 2011. The economics of reseeding on a dairy farm. *Irish Journal of Agricultural & Food Research*, 50(1), p.113.

Sharma, A. K. & Kasana, Æ. R. K. S. Æ. H. S., 2006. Empirical comparisons of feed-forward connectionist and conventional regression models for prediction of first lactation 305-day milk yield in Karan Fries dairy cows. *Neural Computation*, pp.359–365.

Sharma, A. K., Sharma, R. K. K. & Kasana, H. S. S., 2007. Prediction of first lactation 305-day milk yield in Karan Fries dairy cattle using ANN modeling. *Applied Soft Computing*, 7(3), pp.1112–1120. Available at: http://www.sciencedirect.com/science/article/pii/S1568 494606000585 [Accessed October 20, 2014].

Sherchand, L. et al., 1995. Selection of a mathematical model to generate lactation curves using daily milk yields of Holstein cows. *Journal of dairy science*, 78(11), pp.2507–13. Available at: http://www.journalof dairyscience.org/article/S0022-0302(95)76880-1/abstract [Accessed June 3, 2014].

Sikka, L. C., 1950. A study of lactation as affected by heredity and environment. *Journal of Dairy Research*, 17(3), pp.231–52. Available at: http://www.cabdirect.org/abstracts/19510401105.html;jsessionid= F764018286DFA5D33C0626815B853D65.

Silvestre, A. M., Petim-Batista, F. & Colaço, J., 2005. Genetic parameter estimates of portuguese dairy cows for milk, fat, and protein using a spline test-day model. *Journal of dairy science*, 88(3), pp.1225–1230. Available at: http://www.ncbi.nlm.nih.gov/pubmed/15738256.

Silvestre, A. M., Petim-Batista, F. & Colaço, J., 2006. The Accuracy of Seven Mathematical Functions in Modeling Dairy Cattle Lactation Curves Based on Test-Day Records From Varying Sample Schemes. *Journal of Dairy Science*, 89(5), pp.1813–1821. Available at: http://www.ncbi.nlm.nih.gov/pubmed/16606753.

Smith, L. P., 1968. Forecasting annual milk yields. *Agricultural Meteorology*, 5(3), pp.209–214. Available at: http://www.sciencedirect.com/science/article/pii/0002157168900046.

Specht, D. F., 1991. A general regression neural network. *IEEE Transactions on Neural Networks*, 2(6), pp.568–576. Available at: http://ieeexplore.ieee.org/xpl/login.jsp?reload=true&tp=&arnumber=97934&url=http%3A%2F%2Fieeexplore.ieee.org%2Fxpls%2Fabs_all.jsp%3Farnumber%3D97934.

Spiegel, M. R., 1971. *Schaum's Outline of Theory and Problems of Advanced Mathematics for Engineers and Scientists*, McGraw Hill Professional.

Teagasc, 2016. *Dairy Road Map for 2025.* (August), pp.2014–2015. Available at: https://www.teagasc.ie/media/website/publications/2016/Road-map-2025-Dairy.pdf [Accessed August 6, 2017].

Teagasc, 2011. *National Farm Survey Results 2016. Dairy Enterprise.* pp.1–3. Available at: https://www.teagasc.ie/publications/2017/national-farm-survey-results-2016---dairy-enterprise.php [Accessed August 7, 2017].

Thomas, C. V. & DeLorenzo, M. A., 1994. Simulating Individual Cow Milk Yield for Milking Parlor Simulation Models. *Journal of Dairy Science*, 77(5), pp.1285–1295. Available at: http://www.ncbi.nlm.nih.gov/pubmed/8046070 [Accessed December 15, 2015].

Torres, M., Hervás, C. & Amador, F., 2005. Approximating the sheep milk production curve through the use of artificial neural networks and genetic algorithms. *Computers & Operations Research*, 32(10), pp.2653–2670. Available at: http://www.sciencedirect.com/science/article/pii/S0305054804001522.

Upton, J. et al., 2014. A mechanistic model for electricity consumption on dairy farms: definition, validation, and demonstration. *Journal of dairy science*, 97(8), pp.4973–84. Available at: http://www.sciencedirect.com/science/article/pii/S0022030214003993 [Accessed May 6, 2016].

Upton, J. et al., 2015. Investment appraisal of technology innovations on dairy farm electricity consumption. *Journal of dairy science*, 98(2), pp.898–909. Available at: http://www.scopus.com/inward/record.url?eid=2-s2.0-84922278868&partnerID=tZOtx3y1 [Accessed April 7, 2016].

Voyant, C. et al., 2011. Optimization of an artificial neural network dedicated to the multivariate forecasting of daily global radiation. *Energy*, 36(1), pp.348–359. Available at: http://www.sciencedirect.com/science/article/pii/S0360544210005955.

White, I. M., Thompson, R. & Brotherstone, S., 1999. Genetic and environmental smoothing of lactation curves with cubic splines. *Journal of dairy science*, 82(3), pp.632–638. Available at: http://www.ncbi.nlm.nih.gov/pubmed/10194684.

Wilmink, J. B. M., 1987. Adjustment of lactation yield for age at calving in relation to level of production. *Livestock Production Science*, 16(4), pp.321–334. Available at: http://www.sciencedirect.com/science/article/pii/0301622687900029 [Accessed January 12, 2014].

Wong, W. K., Xia, M. & Chu, W. C., 2010. Adaptive neural network model for time-series forecasting. *European Journal of Operational Research*, 207(2), pp.807–816. Available at: http://dx.doi.org/10.1016/j.ejor.2010.05.022.

Wood, P. D. P., 1967. Algebraic Model of the Lactation Curve in Cattle. *Nature*, 216(5111), pp.164–165. Available at: http://www.nature.com/nature/journal/v216/n5111/abs/216164a0.html [Accessed January 12, 2014].

Zar, J. H., 1984. *Biostatistical Analysis* 2nd ed., Englewood Cliffs: Prentice-Hall.

Zarzalejo, L. F., Ramirez, L. & Polo, J., 2005. Artificial intelligence techniques applied to hourly global irradiance estimation from satellite-derived cloud index. In *Energy*. Measurement and Modelling of Solar Radiation and Daylight- Challenges for the 21st Century. pp. 1685–1697. Available at: http://www.sciencedirect.com/science/article/pii/S036054420400249X.

Zhang, F. et al., 2014. Comparative accuracy of lactation curve fitting models in Irish pasture based dairy systems. In *Proceedings from the 31th International Manufacturing Conference*. Cork Ireland, pp. 283–293.

Von Zuben, F. J. & de Andrade Netto, M. L., 1995. Second-order training for recurrent neural networks without teacher-forcing. In *Proceedings of ICNN'95 - International Conference on Neural Networks*. IEEE, pp. 801–806. Available at: http://ieeexplore.ieee.org/document/487520/ [Accessed October 18, 2017].

In: Dairy Farming
Editor: Anke Hertz
ISBN: 978-1-53613-969-3

Chapter 3

Contributions of Life Cycle Assessment to the Sustainability of Milk Production

Laurine S. Carvalho[1], Camila D. Willers[1], Henrique L. Maranduba[2], Sabine Robra[3], José A. Almeida Neto[4] and Luciano B. Rodrigues[2,*]

[1]Federal Institute of Bahia, Brazil
[2]State University of Southwest Bahia, Itapetinga, Brazil
[3]Unit of Environmental Engineering, University of Innsbruck, Austria
[4]State University of Santa Cruz, Ilhéus, Brazil

Abstract

Today, milk and dairy products are an important part of the human diet all over the world. The food processing sector is responsible for negative impacts on the environment, which increase with more intensive production. In this context, the growing interest in sustainable food production has called for the development of methods to increase productivity and simultainously maintain or even reduce the level of natural resources consumption. Life Cycle Assessment (LCA) is a

[*] Corresponding Author Email: rodrigueslb@uesb.edu.br.

method for the evaluation of environmental impacts of products during their life cycle, which permits the identification of critical points in processes and production stages and the description of their environmental as well as resource-related issues. This review focuses on scientific LCA studies conducted on a variety of different milk production systems, including the treatment of milk co-products, different allocation methods, the assessment of environmental impacts caused by fertilizers and agrochemicals in feed grain production and in the different stages of milk production. However, the methodology requires a higher degree of standardization, especially for the analysis of complex agricultural and livestock systems and their various forms and characteristics. Additional studies are necessary, for example on comparable dairy farm systems, in order to construct a wider database. The findings of those studies can help to improve resource use and productivity, and consequently, the environmental performance of the sector. The promotion of more sustainable practices is considered one of the most important contributions of LCA studies on dairy production systems.

Keywords: livestock, milk, dairy cattle, environmental management, impact assessment

1. INTRODUCTION

Milk and dairy products have become a very important part of human diets in many parts of the world.

The average consumption in developed countries has reached 220 liters per capita per year. In developing countries, the consumption of milk and dairy products is growing fast. Between 1968 and today, the annual consumption per capita increased from 28 kg to 45 kg (2017), and based on estimated population growth, increased income and progressing urbanization, will probably reach 66 kg until 2030 (FAO, 2015, Mu et al., 2017).

During recent years, dairy farms have undergone substantial modernization to increase production, like improved animal nutrition and modern technology (Paura & Arhipova, 2016). Genetic improvement of dairy cattle, mechanized and automated feeding and milking systems and humane handling of the milking cows in order to avoid stress are other examples for modernization. The management of pastures and their seasonal fluctuations, veterinary assistance, and feed efficiency also were modernized (Murphy et al., 2014).

The growing food processing sector causes environmental impacts, which contribute to climate change, soil degradation, freshwater consumption and pollution and loss of biodiversity, directly related to more intensive production systems (Noya et al., 2017). In this context, the growing interest in sustainable food production calls for improved agricultural practices to increase productivity, and at the same time reduce or maintain the consumption of resources (Murphy et al., 2017).

The livestock sector is responsible for considerable adverse effects on the environment, due to its emissions from enteric fermentation and manure management and its high demand for natural resources. It is also a significant economic sector and a major employer. To improve the sector's environmental performance, its interactions with the environment have to be identified and analyzed (Willers et al., 2017). Life Cycle Assessment (LCA) is a methodology for the evaluation of the environmental performance of products during their life cycle, with a strong scientific base, internationally recognized and widely used (Willers & Rodrigues, 2014).

1.1. World Cow's Milk Production

The world's key milk producing countries in 2016 are presented in Table 1.

The US ranked first in global milk production. Small dairy farms with less than 100 cows are frequent, but there are also producers with up to 500 lactating cows. Technologically more advanced very large dairy farms with up to 15,000 lactating cows also exist. The country has a strong internal demand for dairy products. The exports amount to 7.8% of the national production, while 4.1% are imported. Principal destinations for North

American dairy products are Mexico, Saudi Arabia, and Asian countries (Zocal, 2017).

A considerable part of India's population is vegetarian and relies on dairy products as a high-quality protein source, which explains the country's high demand of this commodity. India's livestock sector is almost five times larger than the North American equivalent, however, low genetic potential for milk production, insufficient feed management and lack of veterinary assistance for dairy cows result in a very low milk production (Peterson, 2016).

Approximately 80% of India's milk is produced by the unorganized sector. Smallholders with an average of 7 cows are typical for the country's prevailing dairy system, although larger farms with an average of more than 140 lactating cows also exist (Zocal, 2017).

Table 1. Major milk producing countries in 2016

Region	Principal producers (countries or states)	Cattle (million heads)	Milk Production in 2016 (in million metric tons)[2]	Mean productivity per cow (kg/cow/year)
USA	California, Wisconsin, New York, Idaho, Michigan and Pennsylvania	9.2	96.34	10,150
India	Uttar Pradesh, Andhra Pradesh, Gujarat and Punjab	45.9	68	1,446
China	---	12.6	36.02	2,994
Russia	---		30.47	4,029
Brazil	Minas Gerais, Rio Grande do Sul, Paraná	23	22.73	1,525
EU-28*	Germany, France, Netherlands, Poland and Italy [1]	23.508*[1]	151	6,570

Sources: Zocal, 2017; USDA, 2017a; USDA, 2017b.

*Numbers refer to the entire EC (EC28).

China is another major milk-producing country. The size of typical dairy farms varies from 200 to 3,900 lactating cows. According to Ribeiro (2015), China was responsible for increased dairy product prices in international markets, caused by a rapidly growing demand within a short period of time. By 2020, fueled by lower feed prices and improved genetic productivity, China's milk production is expected to have grown sufficiently to reduce its dairy imports (USDA, 2017a).

At the beginning of the 21st century, Russia's dairy sector changed considerably, causing milk production to decrease by 4.5%. The productive dairy cattle population was reduced by 40.7%, while the productivity per cow increased by 61% (Zocal, 2017). In coming years, Russian milk production is expected to shrink by approximately 3% per year. This decline is due to falling prices of Russian dairy products in combination with low investments into the improvement of the genetic potential and management systems of dairy cattle. Another factor affecting Russian milk production is the food industry's preference of cheaper vegetable oil over milk fat. Contrary to global trends, the consumption of fresh milk in Russia is predicted to fall because of high consumer prices (USDA, 2017b).

Although Brazil`s productive cattle population ranks second only behind India, the country is a large importer of dairy products. The size of typical dairy farms ranges from about 20 to more than 300 cows. Brazil's dairy sector is an important part of the country's agribusiness and employs more than 2 million workers (Zocal, 2017). Increased governmental support for the sector and low production costs are the main driving forces behind the production growth in recent years (Sheth, 2018).

1.2. Environmental Aspects and Impacts of Milk Production

During the last 60 years, traditional extensive agricultural production systems have been gradually modernized and intensified to meet the world's growing demand for food (González-García et al., 2013). Thus, the food production sector has become a relevant factor for resources consumption and climate change (Schau & Fet, 2008). Dairy farming has also changed during time, from low input pasture-based production to intensive systems. This modernization process has led to an increased consumption of operating resources like fertilizers, energy, irrigation water, and to the discharge of large amounts of polluting effluents, resulting in an increased environmental burden of the products (Basset-Mens et al., 2009, Claudino & Talamine, 2013). Globally, the agricultural and livestock sector is estimated to produce 18% of the total greenhouse gas emissions (FAO, 2010).

Apart from resource consumption and the residues produced by the sector, soil degradation from overgrazing and constant trampling is detrimental to the soil's capacity to store and supply water and nutrients and consequently, to its fertility (Hamza & Anderson, 2005).

These circumstances raised concerns over the environmental sustainability of the current food production systems, creating a demand for sustainability indicators based on the products' life cycle (Claudino & Talamine, 2013). Consumers are increasingly aware of how products are consumed and also are more and more concerned with environmental and social issues associated with consumption (UNEP, 2002; ERSCP, 2004). To develop more sustainable agricultural systems, researchers and decision-makers need detailed and comprehensive information about the strengths and weaknesses of the various production systems and their environmental impacts (Meier et al., 2015).

Given the economic importance of the sector, the assessment of the relations between production and environmental aspects and impacts has gained considerable importance in the last decades. Every human activity will cause environmental impacts, or modifications of the environment, according to the environmental aspects of that activity. The environmental aspects can be defined as the elements of the activities, products or services of an organization that interact with the environment, and cause environmental impacts, positive as well as negative (ISO 14001, 2015).

Several methodologies have been developed to identify and assess production related environmental burdens. The concept of Life Cycle Assessment (LCA), one example of these methodologies, will be presented in more detail in the following text, including selected research papers on LCAs of milk production systems.

2. Life Cycle Assessment

2.1. Key Concepts

The compliance with environmental laws and standards brings about the need to reduce the negative environmental impacts inherent to products, processes or services. Initially, the focus of environmental management practices was on emission reduction according to the "end-of-pipe" strategy, by implementing treatment systems for effluents and other residues generated by the production process.

Later on, end-of-pipe solutions were perceived as expensive, because they require certain infrastructural and operative measures to adequately treat and dispose of the residues and effluents. To reduce the necessity of expensive treatment facilities, strategies for the reduction of residues were put into place. Re-utilization, recycling (internal or external), as well as the recovery of materials and energy reduce treatment and disposal costs, and therefore can result in considerable economic advantages. Some of these measures are, for example, process optimization to avoid the generation of residues, substitution of input materials by environmentally more friendly alternatives, and transformation of residues into products or co-products. All of these measures potentially reduce environmental impacts, and by integrating new value chains can even generate income.

Approaches considering the whole life cycle of a product, therefore, will assess the complete product system, beginning with the acquisition of raw materials or natural resources, then cover processing, distribution and consumption, and finally disposal. In all life cycle stages, the products that interact with other systems are called open cycles. In order to manufacture a certain product, materials, energy, workforce, technology and financial resources are necessary. Apart from the product, other substances are also emitted to the environment.

The concept of life cycle focuses on a systemic view by combining existing strategies of consumption and production, and thus avoiding a fragmented approach. Using life cycle approaches helps avoid the "displacement problem," that is, incompletely solved problems will be dislocated from one stage of the life cycle to another, from one location to another, from one environment (for example air, water or land) to another, or even from the present to the future (UNEP, 2005).

In this context, Life Cycle Management (LCM) is used to apply environmental sustainable production designs to the complete life cycles of products and services. Life Cycle Assessment (LCA), on the other hand, is a methodology for the systematic evaluation of environmental aspects of a production system, by examining all stages of the respective life cycle (UNEP, 2005).

LCA studies were first conducted in the United States around the beginning of the 1970s, as reaction to the rising awareness that raw materials and energy resources are finite, and evidenced by the first energy crisis with its surging oil prices. These concerns raised interest in methods for the quantification of energy and materials consumption of future projects (Vigon et al., 1993; Santos, 2006).

In the attempt to standardize LCA research, the International Organization for Standardization (ISO) published the family of standards ISO 14040, with ISO 14040 (1997) on general principles of LCA, ISO 14041 (1998) on the definition of goal and scope and the Life Cycle Inventory (LCI), and ISO 14042 (2002) about Life Cycle Inventory Analysis (LCIA), and ISO 14043 (2000), which comprised the interpretation of LCA results.

In 2006, the series was updated and substituted by the following:

- ISO 14040:2006 – Life Cycle Assessment – Principles and Framework.
- ISO 14044:2006 – Life Cycle Assessment – Requirements and Guidelines

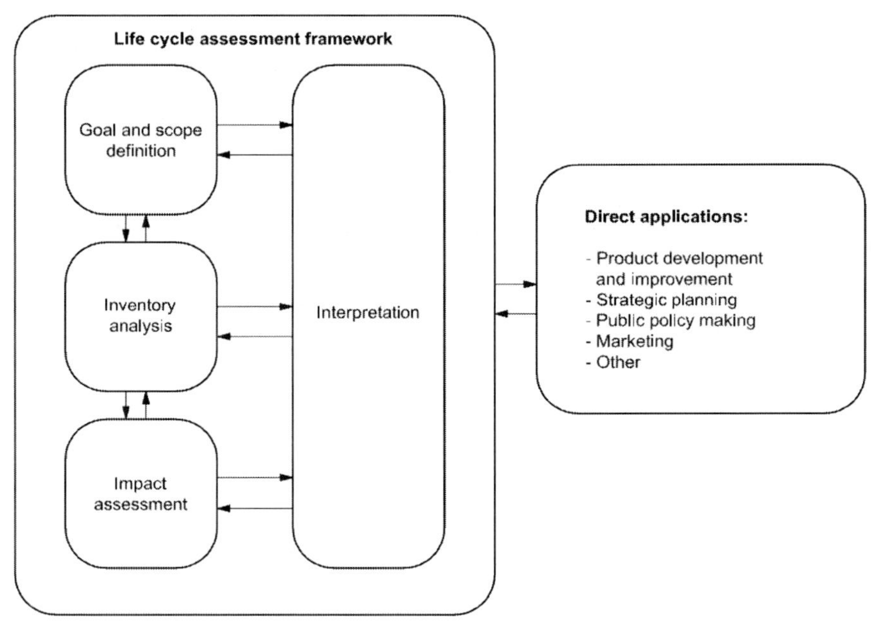

Source: ISO 14040 (2006).

Figure 1. Stages of an LCA.

According to ISO 14040 (2006), LCA is defined as a compilation and evaluation of the inputs, the outputs and the potential environmental impacts of a production system throughout its life cycle. ISO 14040 (2006) also specifies the stages of the LCA (Figure 1).

- Goal definition: the intended application, the motives for the realization of the study and the targeted public are determined;
- Scope definition: Defines the extent of the study, or more specifically, the system boundaries, as well as its function and the functional unit.

A system can have various functions. Their choice depends on the study's goal and scope.

The functional unit defines how the identified functions are quantified, based on the performance characteristics of the product, and serves as a reference to which the inputs and outputs can be related. The system boundary determines the elementary processes that are to be included in the system (ISO 14040, 2006). Example 1 shows the determination of the scope and system boundary of the product. (Figure 2). It also becomes clear that the chosen FU has quantitative as well as qualitative characteristics.

The study in the example evaluates every activity involved in the production of one liter of refrigerated milk, or, in other words, analyzes all the processes related to the production of animal feed, the inputs needed for the herd and for milking, as well as refrigeration of raw milk and its transport to the dairy. Since only the milk production system is under scrutiny, subsequent processing stages, such as pasteurization, the production of milk powder, butter, cheese, yogurt etc., are not included.

The example is a "cradle to gate" study, because it only considers the agricultural production stages (the "cradle") until the delivery of the product to the dairy (the "gate"). Studies "from cradle to grave" include the extraction or production of the resources and also all other stages from processing, distribution, consumption, discharge, to final disposal: Studies "from gate to grave" investigate the intermediate stages of a product from a certain starting point, until its end of life, when it turns into residue or waste.

In Figure 3, the analyzed stages in the exemplary life cycle "from cradle to grave" of milk production are highlighted.

Example 1:
Scope: "cradle to gate" study, the unit process related to milk production in a certain region.
Function: provide protein to human nutrition.
Functional Unit (FU): 1 kg FPCM (1 kg fat and protein corrected milk). In this case, the IDF (2010a) recommends using FPCM as the functional unit, because it allows the comparison of farms with different livestock and feed systems and because the quality of dairy products is determined by the milk fat and protein content.

Figure 2. Defining the Scope, Function and Functional Unit of an LCA.

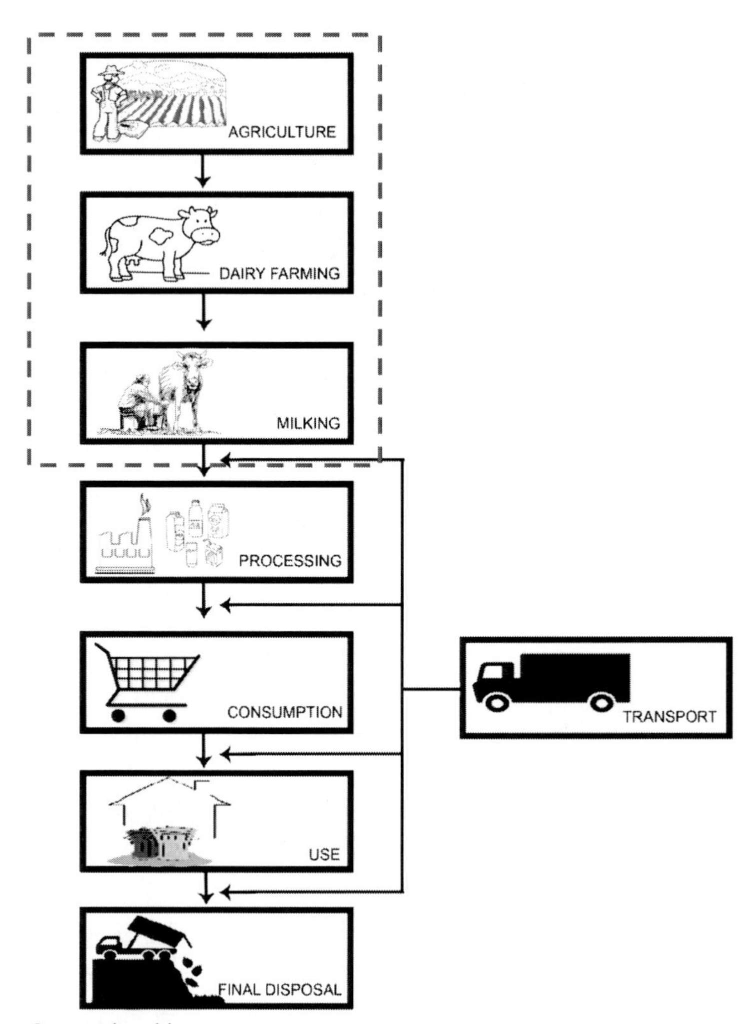

Source: Own authorship.

Figure 3. Life Cycle of milk production.

Inventory Analysis: In this stage of the LCA, data are collected and calculating procedures for the quantification of inputs and outputs of a production system are defined. The inputs comprise all the resources and raw materials used in the system. The outputs include all products and emissions to the atmosphere, to the water or to the soil. These data constitute the starting point for the Life Cycle Impact Assessment (ISO 14040, 2006).

2.1.1. Allocation

During the elaboration of the Life Cycle Inventory, the need of allocation has to be considered, by which the inputs and outputs of a product system are distributed equally among its main product(s) and co-products. An allocation can be necessary when a system generates more than one product, or when recycling is involved (ISO 14040, 2006).

ISO 14044 (2006) recommends to avoid allocation in LCA whenever possible, because it requires more calculations, risking additional uncertainties (Ramirez et al., 2008). Allocation can be avoided by dividing the process into sub-systems or by system expansion (ISO 14044, 2006).

In allocation, the system's inputs and outputs have to be distributed among the various products or functions, so that they reflect the physical relations among them (ISO 14044, 2006), or more specifically, how they change with quantitative modifications of the products or functions that leave the system (Ramirez et al., 2008). When separate physical relations cannot be established or cannot be used as base for allocation, other relations between the inputs/outputs and the products and functions can be used. For example, inputs and outputs can be allocated to the co-products proportionally, according to their economic value (ISO 14044, 2006). The drawback of the economic allocation is its susceptibility to market fluctuations. In order to minimize this effect, the use of mean values of economic values is recommended (Ramirez, 2009).

Feitz et al., (2007) recommend the use of allocation based on physical-chemical properties of the processes and emissions, such as mass, volume, or energy, in order to avoid the above-mentioned errors that economic allocation can cause.

Milk production is not an isolated system because together with it, various co-products are also generated, such as meat, horns, calves, leather etc., the impacts of which need to be distributed, or allocated, among them adequately. Material and energy flows and emissions have to be allocated to the various products with proper documentation and explanation, according to the normalized procedures (ISO 14040, 2006). According to O'Brien et al., (2014), the environmental burden can be distributed between the milk and the meat using the following methods:

1. Milk: no allocation to the meat; the whole environmental burden is allocated to the milk;
2. Mass: the environmental burden of the dairy system is distributed among the co-products according to the mass of sold milk and meat;
3. Economic: The allocation between milk and meat is based on the financial return obtained for the involved products (sale of the cows at the end of their productive life, of the male calves, and of the milk);
4. Protein-based: the environmental burden is allocated to the mass of edible milk and meat protein produced by the system, based on a carcass protein content of 20% (Flysjö et al., 2011);
5. Physical: The dairy system's environmental impacts are allocated based on the nutritional energy necessary for milk and meat production (IDF, 2010a; Doublet et al., 2013);
6. System expansion: the meat of cull cows at the end of their productive life and of the surplus bull calves substitutes meat from beef production systems (Flysjö et al., 2012).

2.1.2. Data Quality Analysis

The previously collected inventory data have to represent the studied product system. However, whenever primary data (collected *in situ*) are not available, secondary data are necessary, which usually come from other studies, technical reports, etc. Even coming from comparable production systems, these data can be a source of uncertainties.

The inventory data possess different degrees of certainty, because some are obtained from primary sources (i.e., were collected in the studied production system, *in situ*), while others are secondary data (i.e., obtained from other sources, like publications technical reports, other studies etc. on similar topics), usually used when primary data are not available. Hence, data quality of both primary and secondary sources depends on aspects like confidence, completeness, temporal and geographical correlation, which, in turn, are affected by the respective sampling procedures applied in every case. In order to assess data quality, an uncertainty analysis can be conducted. Either be based on traditional statistics or else on mathematic simulation methods, the uncertainty analysis estimates the variability of the data by iterations and gives a measure of the uncertainty of the study's results.

A sensitivity analysis is a complementary method to assess data quality. It has to be carried out whenever several different allocation alternatives seem to be appropriate. It consists of systematic procedures to evaluate the consequences of each allocation alternative, or rather, to identify the influence of changes of data and methods on the results of the Life Cycle Impact Assessment (LCIA). The sensitivity analysis determines how uncertain data, allocation methods, or calculation of the category indicators affect the reliability of the final results and conclusions (ISO 14040, 2006, ISO 14044, 2006).

2.1.3. Life Cycle Impact Assessment (LCIA)

In LCIA, the inventory data compiled in the LCI are translated into the specific environmental impacts of the product system. The impact assessment is carried out by classification and characterization of the LCI impact categories of each category indicator and characterization model. The impact categories represent the environmental issues associated to the results of the LCI.

The relative contribution of each of the product system's inputs and outputs is attributed to impact categories and converted into an indicator representing the corresponding potential environmental impact. The results obtained in the selection and classification phase are multiplied by the characterization factor of each substance in each impact category (Figure 4).

When selecting the impact categories, an indicator is chosen in some place of the environmental mechanism (ISO 14047, 2012). Some of the most common impact categories in milk production and their respective indicators are listed in Table 2.

Climate Change: Greenhouse gases (GHG) have a radiative forcing effect (IPCC, 2006a; IDF, 2010b), and therefore are considered the major cause for global warming and climate change (SLADE et al., 2016). Among the greenhouse gases, carbon dioxide (CO_2), nitrous oxide (N_2), and methane (CH_4) are the best-known representatives. Their global warming potential is expressed in kilograms of CO_2 equivalent (CO_2eq). The carbon footprint of a product is an expression of the sum of those gases emitted during its life cycle, within the established system boundaries and for a specific application and a determined quantity of said product (IDF, 2010b; LÉIS, 2013).

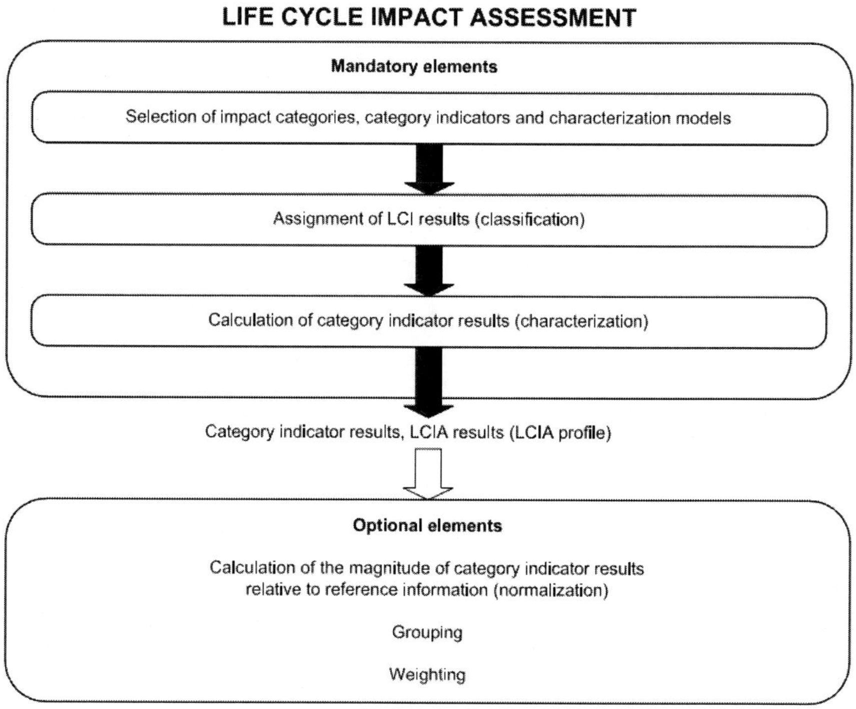

LIFE CYCLE IMPACT ASSESSMENT

Mandatory elements

Selection of impact categories, category indicators and characterization models

↓

Assignment of LCI results (classification)

↓

Calculation of category indicator results (characterization)

↓

Category indicator results, LCIA results (LCIA profile)

↓

Optional elements

Calculation of the magnitude of category indicator results relative to reference information (normalization)

Grouping

Weighting

Source: ISO 14040 (2006).

Figure 4. Elements of the Life Cycle Impact Assessment.

Table 2. Examples of environmental impact categories in milk production LCA

Impact Category*	Characterization factor	Results	Unit
Climate Change	Global Warming Potential (GWP)	Estimation of the greenhouse gases (GHG) emissions	kg CO2-eq
Terrestrial Acidification	Acidification Potential (AP)	Estimation of the emission of acidifying pollutants	kg SO2-eq
Fresh Water Eutrophication	Environmental Persistence (fate) of the emission of P-containing nutrients	Estimation of the nutrients accumulation in freshwater bodies	kg P-eq
Agricultural Land Use	Loss of land as a resource due to use	Estimation of the land area transformed or occupied for a certain time period	m2.year
Fossil Depletion	Fossil resource scarcity	Estimation of fossil fuel consumption, based on the higher heating value (HHV)	kg oil-eq

* ReCiPe midpoint impact indicators method V1.12/Characterization.

According to Yan et al., 2013, most LCA studies concentrate on CO2-eq emissions due to the worldwide concern about global warming.

Terrestrial Acidification: Inorganic substances like sulfates, nitrates and phosphates in the lower atmosphere cause acid rain, because those substances are dissolved by water vapor in the atmosphere. Every plant species has specific requirements concerning soil acidity, and changes can be harmful to certain species (Goedkoop et al., 2009). Acid rain changes the acidity of soils and water bodies and has a damaging effect on forests (Léis, 2013). The indicator of this impact category is expressed in kilograms of sulfur dioxide (kg SO_2eq). Terrestrial Acidification is one of the various impact categories to be considered in an LCA, and its documentation requires an assessment method and a description of the relevant emissions on which the results are based (FAO, 2015b).

Freshwater Eutrophication: The enrichment of aquatic environments with nutrients as a consequence of phosphate and/or nitrous compounds emissions affects the growth pattern in ecosystems. Eutrophication is provoked by excess Nitrogen (N) in the soil (NOx, NHx, NO_{3-}), or by lixiviation of phosphates and nitrates to the water (Boer, 2003; Léis, 2013). The indicator for Eutrophication is expressed either in kilograms P or phosphorous equivalent (kg P_{eq}). Directly or after application of liquid manure or chemical fertilizers on the fields, nutrients (principally phosphorus and nitrogen) can leach and contaminate superficial water bodies. Once in the water, these nutrients are available to the aquatic vegetation and algae, and in excess cause proliferation of the aquatic biomass. After this biomass has died off, its bacterial decomposition is highly oxygen demanding, and quickly can lead to anoxic conditions, causing mortality of fish and other aquatic fauna (FAO, 2015b).

Agricultural Land Use: The environmental relevance of this category in LCA studies is widely accepted (Baan et al., 2012), principally in the agricultural and livestock sectors. According to Alvarenga (2013), the LCA community is aware of the importance of the impacts of land use, and various efforts have been made to improve its evaluation methods. Land use reflects the impacts of occupied areas, either by direct use or by land transformation (Léis, 2013). The agricultural stages of production processes usually require the largest area in relation to the amounts of finished product. The category indicator of Land Use is expressed in m^2 per year (Guinée et al., 2001).

Fossil Depletion: The term "fossil fuel" refers to a group of resources containing hydrocarbons, and includes volatile materials, like methane, and non-volatile materials, like coal. The formation of fossil fuels dates to the Permian period (230 million years ago), the Jurassic period (150 million years ago), and the Cretaceous Period (144-65 million years ago). During those periods, vast amounts of plankton and other organisms proliferated in an environment of high temperatures and then were deposited on the bottom of oceans and expansive lakes, transforming into large crude oil and gas reservoirs. Approximately 1% of the original deposits can still be found in quantities worth exploring (Goedkoop et al., 2009). During extraction, crude oil and natural gas flow freely from the well, up to a certain point. After that, crude oil extraction requires an increasing amount of raw materials, supplies and energy, resulting in higher extraction costs. Increasing scarcity of fossil fuels stimulates the production of new energy sources, also known as new non-conventional energy resources, in order to meet the demand. Among the non-conventional fossil fuel sources are tar sands, as well as the nuclear fuel Uranium (in its enriched form ^{235}U). The category "Fossil Depletion" is focused on the substitution of conventional by non-conventional fossil fuels, principally because in most cases, these are expected in LCA due to their importance on a global scale. Since non-conventional fossil resources generally are more energy intensive and more expensive to produce than their conventional counterparts, their production is only economically viable once conventional fuel prices have reached a certain level (Goedkoop et al., 2009).

Interpretation of the Results: In this stage, the findings of the Inventory Analysis and of the Impact Assessment are interpreted together, conclusions drawn, limitations identified and recommendations given to the target audience; it should however be noted that this is an independent stage.

2.2. Life Cycle Assessment in Dairy Farming

A large number of scientific LCA studies exist on the analysis of environmental impacts of milk production, regarding a wide variety of aspects (Table 3). Most of the cited publications address different milk production systems (extensive, intensive, semi-extensive, semi-intensive), the production of co-products of milk, as well as the question of allocation, fertilizers and agrochemicals in feed grain production, different stages of milk production, among others. Other LCA studies concentrated on the identification of production stages causing the most relevant environmental impacts, or to compare different milk production methods, such as conventional and organic production.

Table 3. Scientific publications on LCA in dairy farming

Year	Region	Authors	Scope	Production System	Functional Unit	Allocation Method	Impact Categories	Impact Assessment Method	Scenarios	Statistical Analysis	UA	SA
2018	Italy	Zucali et al.	CFG CAM	Intensive	kg FPMC	F	GWP, AP, EP, EU	CML – IA, CED	BASE, HAYa, SILAGEb, PROTEINc	No	No	No
2017	Ireland	Murphy et al.	CFG	Semi-intensive	kg FPMC	E	WF	AQUASTAT, TEAGASC, WSI, WTA	No	No	No	No
2017	Northeast of Spain	Noya et al.	CFG	Intensive	kg FPMC	E; M	GWP, AP, EP, WD, FD	ReCiPe	A; B; C	No	No	No
2017	Ethiopia	Woldegebriel et al.	CFG	Extensive; Semi intensive; Intensive	CU; MPC	E	LU, EU, GWP	Ecoinvent 2.2 "Tier 2" (IPCC, 2006b)	LS PU RS	ANOVA	No	No
2017	Belgic France Germany Ireland Luxemburg Netherlands	Mu et al.	CFG	Semi intensive, Intensive	kg FPMC	E	LU, EU, CH, EP, AP	ReCiPe Dairyman (2010)	No	P and S correlations, Regression Analysis	No	No

Table 3. (Continued)

Year	Region	Authors	Scope	Production System	Functional Unit	Allocation Method	Impact Categories	Impact Assessment Method	Scenarios	Statistical Analysis	UA	SA
2016	Italy	Bacenetti et al.,	CFG	Intensive	kg FPCM	E, B	GWP, AP, EP, PO, EU	ReCiPe	BS, TDM, AD	No	No	Yes
2015	Germany	Kiefer et al.,	CFG	Extensive	kg FPCM	S, F, E	CF	IDF IPCC (2006a)	No	SPSS Software, Regression Analysis	No	No
2014	Ireland	O'Brien et al.,	CFG	Extensive	kg FPCM	E	CF	GHG Model PAS 2050	No	SAS 2008	No	No
2014	Australia	Gollnow et al.,	CFG	Semi intensive	kg FPCM	F	CF	IPCC (2006a) DCCEE	No	No	No	No
2014	Brazil	Léis et al.,	CFG	Extensive, Semi intensive, Intensive	kg ECM	E, M	CF	IPCC (2006a)	MMS DL	No	Monte Carlo	Yes
2014	Iran	Daneshi et al.,	CFG	Extensive	kg FPCM	E, BF	CF	IPCC (2006a)	10-12% of the average CF/kg FPCM	No	No	No
2014	Belgic	Meul et al.,	CFG	Intensive	kg FPCM	E	GWP, AP, EP, EU, LU	MOTIFS	No	MLRA	No	No

Year	Region	Authors	Scope	Production System	Functional Unit	Allocation Method	Impact Categories	Impact Assessment Method	Scenarios	Statistical Analysis	UA	SA
2013	Ireland	Yan et al.,	CFG	Extensive	kg ECM	E	CF	IPCC (2000) GWP	No	No	No	Yes
2012	Canada	Geough et al.,	CFG	Extensive Intensive	kg FPCM	E; F	GHG	Holos	(1) (2)	No	No	No

UA: Uncertainty analysis; SA: Sensitive analysis. Scope: CFG: Cradle to Farm Gate; CAM: Cradle to the Animal's Mouth; CG Cradle to Grave; CU: 1 adult cattle unit; MPC:1 kg of milk produced by a cow; Scenarios: BASE: baseline scenario; HAY: the entire farm land is dedicated to permanent crops (grass and alfalfa) for hay production; SILAGE: most of the farm land is used for crop cultivation for silage; PROTEIN: the cropping system is managed to maximize protein production from home-grown feed. A: Using alfalfa in animal feed (major impacts); B: Replacement of alfalfa by maize silage; C: Replacement of alfalfa by grass silage. LS: large-scale; PU: (peri-)urban; RS: rural systems. MMS: Manure Management System; DL: Dry Lot; (1) IDF default, where the allocation factor was determined using the default value of 0.025 kg of meat/kg of milk to yield an allocation of 14.4% to meat and 85.6% to milk; (2): IDF specific, where the allocation factor was determined using the meat and milk yield values generated by the simulated farm (R = 0.046), yielding an allocation of 26.6% to meat and 73.4% to milk. Impact Categories: WF: Water Footprint; GWP: Global Warming Potential; AP: Acidification Potential, EP: Eutrophication Potential; CF: Carbon Footprint; EU: Energy Use; LU: Land Use; FD: Fossil Depletion; WD: Water Depletion; PO: Photochemical Oxidation. Impact Assessment Method: AQUASTAT (NEW et al., 2002); TEAGASC: Teagasc advisory database - Teagasc Grass Calculator (TEAGASC, 2011); WSI: Water Stress Index (PFISTER et al., 2009); WTA: Water-To-Availability (FRISCHKNECHT et al., 2006); CML–IA: CML-IA baseline 3.01 method; CED: Cumulative Energy Demand 1.08 method; IPCC: IPCC (2006a); IDF: International Dairy Federation; IPCC: Intergovernmental Panel On Climate Change; PAS 2050: BSI (2011); DCCEE: (DCCEE, 2010a, b); MOTIFS: Monitoring Tool for Integrated Farm Sustainability (MEUL et al., 2008); GWP: Global warming potential. Statistical Analysis: SPSS: Software was used for statistical analysis; SAS 2008: SAS user guide version 9.1.3 (SAS, 2008); MLR: Multiple Linear Regression Analysis; P and S correlations - Pearson and Spearman correlations; BS; TDM; AD: The environmental performances of a conventional intensive dairy farm in Northern Italy (BS: baseline scenario) were compared with the results obtained: from the introduction of the third daily milking (TDM) and from the adoption of anaerobic digestion (AD) of animal slurry in a consortium AD plant (BACENETTI et al., 2016).

Table 3 shows that concerning the system boundary, studies from "cradle to gate" of the dairy farm are the most common, since 78% - 83% of the dairy sector's emissions in Europe, North America and Oceania are generated by the unit process in this system boundary (FAO, 2010). Another common factor in the majority of the papers compiled in Table 3 is the economic allocation of the co-products milk and meat. However, Pelletier & Tyedmers (2011) argue that the economic allocation is not appropriate because environmental concerns are not reflected by market prices. O'Brien et al., (2014) point out that, depending on the chosen allocation method for milk and meat, the relative difference of the carbon footprint of confined and pasture-based systems can vary from 3% to 22%.

Many papers concentrated on the same impact categories, such as Climate Change, sometimes also referred to as "Carbon Footprint" or "Greenhouse Gases Emission," depending on the respective impact assessment method used, because of the considerable contribution of agricultural and livestock activities to GHG emissions. In this context, GHG emissions of agricultural activities and those associated to land use change were estimated to amount to 30% of the global anthropogenic emissions in 2010 (Slade et al., 2016). Agricultural soils are a considerable source for nitrous oxide (N_2O) emissions due to the application of nitrogen fertilizers, biological nitrogen fixation, spreading of manure and incorporation of harvest residues, while burning the latter causes methane (CH_4), nitrous oxide (NOx) and carbon monoxide (CO) emissions. In 2005, N_2O emissions amounted to 476 Gg, especially due to animal feces and urine in the pastures (46%), and indirect emissions from fertilizers and soil (32%) (Brasil 2010; Willers et al., 2017).

However, the study of only one impact category in LCA is not recommended, because it does not comply with the principles of the methodology, which the objective is to achieve a comprehensive view of the life cycle, and therefore requires the inclusion of a variety of environmental impacts into the study (Baldini et al., 2017). Other impact categories frequently present in the mentioned papers are Acidification, Eutrophication, Land Use and Abiotic Resource Depletion.

The compiled papers in Table 3 also reflect the importance of LCA based results for milk production and its environmental damages. However, one shortcoming of using scientific LCA studies for reference or comparison purposes lies in the widely differing way the studies are conducted. Although LCA can be used to compare the environmental performance of manufactured products, the same can be quite misleading in case of agricultural products (Beauchemin & Mcgeough, 2013).

An LCA of dairy farming has to consider a variety of different factors, depending on how the study is conducted, the geographical location and available equipment, among others. One of the factors is the choice of the functional unit, which could be a liter, a kilogram or a metric ton of either energy corrected milk (ECM), or fat and protein corrected milk (FPCM), as used by the majority of the studies compiled in Table 3. The scope of the study can also vary: some studies assess the production only to the farm gate, while others include milk processing, and still others use a wider approach and also include the consumption stage in their assessment. Transportation is another factor that varies among the studies, as it is included in some, but not in others. The complexity of the studied agricultural and livestock systems and the numerous possible differences make it difficult to compare the environmental performances of those systems. However, the potential of the methodology is not diminished by those factors, and satisfying results in identifying environmental impacts and ways to mitigate them can still be obtained. Therefore, and because animal production systems are so closely connected to the environment, LCA is still useful tool to assess the production system's environmental impacts.

The results of Life Cycle Impact Assessments of different production systems are shown in Table 4.

The assessment of the environmental impacts per milk unit is hampered by the impossibility to clearly identify the relationship between farming intensity and environmental performance, despite some important differences in terms of farm intensification level, management and structural characteristics (Bava et al., 2014). As shown in Table 4, the

Table 4. Environmental impacts of different milk production systems

Environmental Impact Categories	Intensity level				
	Low		Medium	High	
Climate Change (kg CO2 eq./kg FPMC)	0.8742	1.253	1.273	1.321	1.263
Terrestrial Acidification (kg SO2 eq./kg FPMC)	6.92	13.93	16.23	27.81	16.03
Eutrophication (kg PO4 eq./kg FPMC)	3.42	6.863	7.213	37.11	7.593
Land Use (m2/year)	0.7282	0.973	1.023	-	0.893

Sources: [1] Noya et al., (2017); [2] O'Brien et al., (2012); [3] Bava et al., (2014).

intensification of production systems, with increased consumption of resources, fertilizers, agrochemicals and operating materials for feed production, among others, can increase pressure on environmental systems. According to Capper et al., (2008), a general increase in productivity might affect the environmental sustainability of milk. However, the effects of the intensification of milk production on the environment are not yet completely explained (Bava et al., 2014). One effect of different intensification levels is related to the respective scale of consideration. On a global scale, intensification can be seen as positive. On the local scale however, the environmental impacts tend to increase with increasing intensification (Seó et al., 2017; Bava et al., 2014).

In view of population growth and the rising demand of dairy products, one concern of the researchers is how to realize the intensification of production systems and, at the same time, mitigate or avoid the environmental damages caused by those systems. Sustainable intensification of agriculture can be interpreted in different ways, from intensification in terms of increased production and therefore reduced relative environmental impacts, to more efficient resource utilization in order to provide means for a lasting subsistence for the farmers (Cook et al., 2015; Woldgebriel et al., 2017).

The intensification of pasture-based production systems can reduce impacts associated to climate change, and reduce the consumption of synthetic fertilizers and concentrated feeds. This is a consequence of increased pasture productivity and quality due to improved management, for example, by division of pastures. Managed pastures can also lead to higher amounts of sequestered carbon by photosynthesis and thus, mitigate the effects of enteric fermentation. On a global scale, the best cost-benefit relation of mitigating environmental impacts is seen in increased efforts in regions with low productivity (Seó et al., 2017).

In a decision support system, an LCA on farm level can cover varying environmental impacts among farms with a specific production system. By identifying the best farm-specific strategies, farmers can be provided with targeted advice to optimize the environmental performance of livestock farms (Meul et al., 2014).

CONCLUSION

Agricultural, and more so, milk production systems, are complex and vary widely. Accordingly, studies of their life cycles use varying approaches and procedures. In spite of those difficulties, LCA is a suitable methodology for the identification of critical points in the various stages of a production system, by uncovering its environmental as well as resource-related issues. Once these issues are identified, measures can be proposed and implemented to improve the system's environmental performance. The methodology also allows the comparison of different production systems, considering various intensification levels.

However, much research is required to develop more standardized approaches to the methodology, which specifically address the complexity of agricultural and livestock systems and their various forms and characteristics. More studies need to be performed, for example in comparable dairy farm systems, to construct a wider database. The results of those studies can, in short and medium term, lead to direct improvements in resource use and to higher productivity, resulting in improved environmental performances of the production systems. Therefore, the promotion of progress towards more sustainable practices is regarded one of the most important contributions of LCA studies on dairy production systems.

These efforts are not only justified by the importance of the dairy sector for human nutrition. Reducing environmental impacts also help maintain a healthy and agreeable environment on local and global scales.

REFERENCES

Alvarenga, R. A. F. (2013). *Environmental sustainability of biobased products: new assessment methods and case studies*. Thesis (Doctorate in Applied Biological Sciences: Environmental Technology) - Ghent University.

Baan, L., Alkemade, R., Koellner, T. (2013). Land use impacts on biodiversity in LCA: a global approach. *The International Journal of Life Cycle Assessment*, 18, 1216–1230.

Bacenetti, J., Bava, L., Zucali, M., Lovarelli, D., Sandrucci, A., Tamburini, A., Fiala, M. (2016). Anaerobic digestion and milking frequency as mitigation strategies of the environmental burden in the milk production system. *Science of the Total Environment*, 539, 450–459.

Baldini, C., Gardoni, D., Guarino, M. (2017). A critical review of the recent evolution of Life Cycle Assessment applied to milk production. *Journal of Cleaner Production*, 140, 421-435.

Basset-Mens, C., Ledgard, S., Boyes, M. (2009). Eco-efficiency of intensification scenarios for milk production in New Zealand. *Ecological Economics*, 68, 1615-1625.

Bava, L., Sandrucci, A., Zucali, M., Guerci, M., Tamburini, A. (2014). How can farming intensification affect the environmental impact of milk production? *Journal of Dairy Science*. 97, 4579-4593.

Beauchemin, K. A., McGeough, E. J. (2013). Life Cycle Assessment in Ruminant Production. Sustainable animal agriculture. In. Kebreab, E (Ed.), *Sustainable animal agriculture*, DOI: 10.1079/9781780640426.0000.

Boer, I. J. M. de (2003). Environmental impact assessment of conventional and organic milk production. *Livestock Production Science.* 80(1-2): 69-77.

Brasil. Ministry of Science Technology and Innovation (2010). Emissões de metano por fermentação entérica e manejo de dejetos de animais. In *Segundo inventário brasileiro de emissões antrópicas de gases de efeito estufa* [Emissions of methane by enteric fermentation and management of animal waste. In *Second Brazilian inventory of anthropogenic emissions of greenhouse gases*]. Brasília, DF. [in Portuguese.

BSI - British Standards Institute (2011). *PAS 2050:2011—specification for the assessment of life cycle greenhouse gas emissions of goods and services*, London.

Capper, J. L., Castañeda-Gutiérrez, E., Cady, R. A., Bauman., D. E. (2008). The environmental impact of recombinant bovine somatotropin (rbST) use in dairy production. *Proceedings of the National Academy of Sciences*, 105: 9668-9673.

Claudino, E. S., Talamini, E. (2013). Análise do Ciclo de Vida (ACV) aplicada ao agronegócio – Uma revisão de literatura. *Revista Brasileira de Engenharia Agrícola e Ambiental* [Life Cycle Analysis (LCA) applied to agribusiness - A literature review. *Brazilian Journal of Agricultural and Environmental Engineering*], 17, 77–85. [in Portuguese].

Cook, S., Silici, L., Adolph, B., Walker, S. (2015). *Sustainable intensification revisited.* IIED Issue Paper, IIED, London. ISBN: 978-1-78431-185-8.

Dairyman (2010) *A project in the INTERREG IVB program co-funded by the European Regional Development Fund.* http://www. interregdairyman.eu/en/dairyman.htm.

Daneshi, A., Esmaili-Sari, A., Daneshi, M., Baumann, H. (2014). Greenhouse gas emissions of packaged fluid milk production in Tehran. *Journal of Cleaner Production*, 80, 150-158.

DCCEE - Department of Climate Change and Energy Efficiency. (2010a). National inventory report 2008 (pp. 203e282). In *The Australian government submission to the UN framework convention on climate change,* May (Vol. 1). Canberra, Australia.

DCCEE - Department of Climate Change and Energy Efficiency (2010b). *National greenhouse accounts (NGA) factors* (pp. 10e34). Canberra, Australia.

Doublet, G., Jungbluth, N., Flury, K., Stucki, M., (2013) Life cycle assessment of Romanian beef and dairy products. *SENSE - Harmonised Environmental Sustainability in the European food and drink chain, Seventh Framework Programme: Project no. 288974.* Funded by EC. Deliverable D 2.1 ESU-services Ltd.: Zürich. Retrieved from http://www.esu-services.ch/projects/lcafood/sense/.

ERSCP - European Roundtable on Sustainable Consumption and Production. (2004). *Meeting consumer demand for sustainable products.* Summary of the workshop. Bilbao.

FAO - Food and Agriculture Organization of the United Nations. (2015a). *World agriculture: towards 2015/2030. Long-term Perspectives: The outlook for agriculture.* Available at: http://www.fao.org/docrep/004/Y3557E/y3557e00.htm#TopOfPage.

FAO - Food and Agriculture Organization of the United Nations. (2015b). *Environmental performance of large ruminant supply chains: Guidelines for assessment.* Available at http://www.fao.org/3/a-av152e.pdf.

FAO - Food and Agriculture Organization of the United Nations. (2010). Greenhouse gas emissions from the dairy sector: *A life cycle assessment. Animal Production and Health Division,* Rome, Italy. Available at: http://www.fao.org/docrep/012/k7930e/k7930e00.pdf.

Feitz, A. J., Lundie, S., Dennien, G., Morain, M., Jones, M. (2007). Generation of an Industry-Specific Physico-Chemical Allocation Matrix: Application in the Dairy Industry and Implications for Systems Analysis. *The International Journal of Life Cycle Assessment*, 12, 119-117.

Flysjö, A., Cederberg, C., Henriksson, M., Ledgard, S. (2012). The interaction between milk and beef production and emissions from land use change – critical considerations in life cycle assessment and carbon footprint studies of milk. *Journal of Cleaner Production*, 28, 134-142.

Flysjö A., Cederberg C., Henriksson M., Ledgard S. (2011). How does co-product handling affect the Carbon Footprint of milk? – Case study of milk production in New Zealand and Sweden. *The International Journal of Life Cycle Assessment*, 16, 420-430.

Frischknecht, R., Steiner, R., Braunschweig, A., Egli, N., Hildesheimer, G., (2006). *Swiss Ecological Scarcity Method*: the New Version 2006. Berne, Switzerland.

Geough, E. J., Little, S. M., Janzen, H. H., Mcallister, T. A., Mcginn, S. M., Beauchemin, K. A. (2012). Life-cycle assessment of greenhouse gas emissions from dairy production in eastern Canada: a case study. *Journal of Dairy Science*, 95, 5164 -5175.

Goedkoop, M. J., Heijungs, R., Huijbregts, M. A. J., De Schryver, A. M., Struijs, J., Van Zelm, R. 2009. ReCiPe 2008: *A life cycle impact assessment method which comprises harmonised category indicators at the midpoint and the endpoint level;* First edition Report I: Characterisation. 6 January 2009, http://www.lcia-recipe.net.

Gollnow, S., Lundie, S., Moore, A. D., Mclaren, J., Van Buuren, N., Stahle, P., Christie, K., Thylmann, D., Rehl, T. (2014). Carbon footprint of milk production from dairy cows in Australia. *International Dairy Journal*, 37, 31-38.

González-García, S., Castanheira, E. G.; Dias A. C.; Arroja, L. (2013). Environmental life cycle assessment of a dairy product: The yoghurt. *The International Journal of Life Cycle Assessment* 18, 796 - 811.

Guinée, J. B., Gorrée, M., Heijungs, R., Huppes, G., Kleijn, R., Koning, A. De, Oers, L. Van., Wegener Sleeswijk, A., Suh, S., Udo De Haes, H. A., Bruijn, H. De, Duin, R. Van, Huijbregts, M. A. J. (2001). *Life Cycle Assessment: An operational guide to the ISO standards*. Part 1. https://media.leidenuniv.nl/legacy/new-dutch-lca-guide-part-1.pdf.

Hamza, M. A., Anderson, W. K. (2005). Soil compaction in cropping systems: A review of the nature, causes and possible solutions. *Soil & Tillage Research* 82(2):121–145.

IDF – International Dairy Federation. (2010a). A Common Carbon Footprint approach for Dairy, The IDF Guide to Standard Life cycle. *Bulletin of International Dairy Federation Report*. 445.

IDF – International Dairy Federation. (2010b). The World Dairy Situation 2010. *Bulletin of International Dairy Federation Report*. 446.

IPCC- Intergovernmental Panel on Climate Change. (2000). *IPCC Good Practice Guidance and Uncertainty Management in National Greenhouse Gas Inventories*. Institute for Global Environmental Strategies, Kanagawa, Japan. http://www.ipcc-nggip.iges.or.jp/public/gp/english.

IPCC - Intergovernmental Panel on Climate Change. (2006a). *Guidelines for National Greenhouse Gas Inventories. Agriculture, Forestry and Other Land Use - Emissions from Livestock and Manure Management*, v. 4, Ch. 10.

IPCC- Intergovernmental Panel on Climate Change. (2006b). *IPCC Guidelines for National Greenhouse Gas Inventories*. Prepared by the National Greenhouse Gas Inventories Programme, IGES, Hayama.

ISO - International Organization for Standardization - ISO 14001. (2015). *Environmental Management Environmental management systems* — Requirements with guidance for use.

ISO - International Organization for Standardization - ISO 14040. (2006). *Environmental Management – Life Cycle Assessment – Principles and Framework*.

ISO - International Organization for Standardization - ISO 14044. (2006). *Environmental Management – Life Cycle Assessment – Requirements and guidelines*.

ISO - International Organization for Standardization - ISO 14047. (2012). *Environmental management - Life cycle assessment* - Illustrative examples on how to apply ISO 14044 to impact assessment situations.

Kiefer, L. R., Menzel, F., Bahrs, E. (2015). Integration of ecosystem services into the carbon footprint of milk of South German dairy farms. *Journal of Environmental Management*, 152, 11-18.

Léis, C. M. de. (2013). *Desempenho ambiental de três sistemas de produção de leite no sul do Brasil pela abordagem da Avaliação do*

Ciclo de Vida. Tese (Doutorado). Universidade Federal de Santa Catarina - Programa de Pós-Graduação em Engenharia Ambiental. Florianópolis [*Environmental performance of three milk production systems in southern Brazil by the Life Cycle Assessment approach.* Thesis (Doctorate). Federal University of Santa Catarina - Post-Graduation Program in Environmental Engineering. Florianópolis]. [In Portuguese].

Léis, C. M., Cherubini, E., Ruviaro, C. F., Silva, V. P., Lampert, V. N., Spies, A., Soares, S. R. (2014). Carbon footprint of milk production in Brazil: a comparative case study. *The International Journal of Life Cycle Assessment,* 20, 46-60.

Meier, M. S., Stoessel, F., Jungbluth, N., Juraske, R., Schader, C., Stolze, M. (2015). Environmental impacts of organic and conventional agricultural products - Are the differences captured by life cycle assessment? *Journal of Environmental Management*, 149, 193-208.

Meul, M., Van Middelaar, C. E., Boer I. J. M. de, Van Passel, S., Fremaut, D., Haesaert, G. (2014). Potential of life cycle assessment to support environmental decision making at commercial dairy farms. *Agricultural Systems*, 131, 105–115.

Meul, M., Van Passel, S., Nevens, F., Dessein, J., Rogge, E., Mulier, A., Van Hauwermeiren, A. (2008). MOTIFS: a monitoring tool for integrated farm sustainability. *Agronomy for Sustainable Development,* 28, 321–332.

Mu, W., van Middelaar, C. E., Bloemhof, J. M., Engel, B., Boer, I. J. M. de, (2017). Benchmarking the environmental performance of specialized milk production systems: selection of a set of indicators. *Ecological Indicators*, 72, 91–98.

Murphy, E., Boer, I. J. M. de, van Middelaar, C. E., Holden, N. M., Shalloo, L., Curran, T. P., Upton, J. (2017). Water footprinting of dairy farming in Ireland. *Journal of Cleaner Production*, 140(Part 2): 547-555.

Murphy, M. D., O'Mahony, M. J., Shalloo, L., French, P., Upton, J., (2014). Comparison of modeling techniques for milk-production forecasting. *Journal of Dairy Science*, 97, 3352–3363.

New, M., Lister, D., Hulme, M., Makin, I., (2002). A high-resolution data set of Surface climate over global land areas. *Climate Research*, 21, 1e25.

Noya, I., González-García, S., Berzosa, J., Baucells, F., Feijoo, G., Moreira, M. T., (2018). Environmental and water sustainability of milk production in Northeast Spain. *Science of the Total Environment*, 616-617: 1317-1329.

O’Brien, D., Brennan, P., Humphreys, J., Ruane, E., Shalloo, L. (2014). An appraisal of carbon footprint of milk from comercial grass-based dairy farms in Ireland according to a certified Life Cycle Assessment methodology. *The International Journal of Life Cycle Assessment*, 19, 1469–1481.

O'Brien, D., Shalloo, L., Patton, J., Buckley, F., Grainger, C., Wallace, M. (2012). A life cycle assessment of seasonal grass-based and confinement dairy farms. *Agricultural Systems*, 107, 33–46.

Paura, L., Arhipova, I., (2016). Analysis of the milk production and milk price in Latvia. *Procedia Economics and Finance*, 39, 39–43.

Pelletier, N., Tyedmers, P. (2011). An ecological economic critique of the use of market information in life cycle assessment research. *Journal of Industrial Ecology. 15*, 342–354.

Peterson, C. (2016). How One Farmer Built a Dairy Dynasty in India. *Forbes*. Cargill Voice.

Pfister, S., Koehler, A., Hellweg, S., (2009). Assessing the environmental impacts of freshwater consumption in LCA. *Environmental Science & Technology,* 43, 4098e4104.

Ramirez, P. K. S. (2009). *Análise de Métodos de Alocação Utilizados em Avaliação do Ciclo de Vida.* Dissertação [*Analysis of Allocation Methods Used in Life Cycle Assessment.* Dissertation] (Mestrado em Engenharia Ambiental) – Universidade Federal de Santa Catarina, Florianópolis. [In Portuguese].

Ramirez, P. K. S., Soares, S. R., Souza, D. M. de, Silva Junior, V. M. da (2008). Allocation Methods in Life Cycle Assessment: A Critical Review. *15th CIRP International Conference on Life Cycle Engineering: Conference Proceedings*. LCE.

Ribeiro, R. (2015). A China e sua importância no mercado de leite [China and its importance in the milk market]. *Animal Business.* 24, 43-45. [In Portuguese].

SANTOS, L. M. M. (2006). *Avaliação ambiental de processos industriais.* 2 ed. São Paulo: Signus Editora [*Environmental assessment of industrial processes*], 130p. [In Portuguese].

SAS - Statistical Analysis Systems. (2008). *SAS user guide version 9.1.3.* Institute Inc., Cary.

Schau, E. M., Fet, A. M. (2008). LCA studies of food products as background for environmental product declarations. *The International Journal of Life Cycle Assessment*, 13(3): 255–264.

Seó, H. L. S., Machado Filho, L. C. P., Ruviaro, C. F., Léis, C. M. de. (2017). Avaliação do Ciclo de Vida na bovinocultura leiteira e as oportunidades ao Brasil [Life Cycle Assessment in dairy cattle and opportunities in Brazil]. *Eng Sanit Ambient.* 22, 221-237. [In Portuguese].

Sheth, K. (2018). *Top Milk Producing Countries in the World.* https://www.worldatlas.com/articles/top-cows-milk-producing-countries-in-the-world.html.

Slade, E. M., Riutta, T., Roslin, T., Tuomisto, H.L. (2016). The role of dung beetles in reducing greenhouse gas emissions from cattle farming. *Scientific Reports*, 6, 18140. DOI: 10.1038/srep18140.

TEAGASC (2011). *The Grass Calculator Teagasc, Fermoy, Ireland.* http://www.agresearch.teagasc.ie/moorepark/.

UNEP - United Nations Environment Programme. (2002). *Tracking process: implementing sustainable consumption polices.* ISBN: 92-807-2205-0. http://www.unep.fr/shared/publications/pdf/3083-Tracking Process1.pdf.

UNEP - United Nations Environment Programme. (2005). *Life Cycle Initiative. Life Cycle Approaches: The road from analysis to practice.* Division of Technology, Industry and Economics (DTIE). https://www.lifecycleinitiative.org/wp-content/uploads/2012/12/2005%20-%20LCA.pdf.

USDA - U.S. Department of Agriculture. (2017a). Foreign Agricultural Service. Global Agricultural Information Network. China - Peoples Republic of Food and Agricultural Import Regulations and Standards – Narrative FAIRS Country Report. *GAIN Report Number: CH16069*.

USDA - U.S. Department of Agriculture. (2017b). Foreign Agricultural Service. Office of Global Analysis. *Dairy: World Markets and Trade.* Available at: https://apps.fas.usda.gov/psdonline/circulars/dairy.pdf.

Vigon, B. W., Tolle, D. A., Cornaby, B. W., Latham, H. C., Harisson, C. L., Boguski, T. L., Hunt, R. G., Sellers, J. D. (1993). *Life-Cycle Assessment: Inventory Guidelines and Principles.* United States Environmental Protection Agency.

Willers, C. D., Maranduba, H. L., de Almeida Neto, J. A., Rodrigues, L. B. (2017). Environmental Impact assessment of a semi-intensive beef cattle production in Brazil's Northeast. *The International Journal of Life Cycle Assessment,* 22, 516-524.

Willers, C. D., Rodrigues, L. B. (2014). A critical evaluation of Brazilian life cycle assessment studies. *The International Journal of Life Cycle Assessment,* 19, 144-152.

Woldegebriel, D., Udo, H., Viets, T., van der Harstb, E., Potting, J. (2017). Environmental impact of milk production across an intensification gradient in Ethiopia. *Livestock Science*, 206, 28–36.

Yan, M. J., Humphreys, J., Holden, N. H. (2013). The carbon footprint of pasture-based milk production: Can white clover make a difference? *Journal of Dairy Science*, 96, 857–865.

Zocal, R. (2017). *Dez países top no leite* [*Ten top countries in milk*]. Balde Branco. http://www.baldebranco.com.br/dez-paises top-no-leite [In Portuguese].

Zucali, M., Bacenetti, J., Tamburini, A., Nonini, L., Sandrucci, A., Bava, L. (2018). Environmental impact assessment of different cropping systems of home-grown feed for milk production. *Journal of Cleaner Production*, 172, 3734-3746.

INDEX

A

B

C

D

E

F

G

H

I

O

P

Q

R

S

T

U

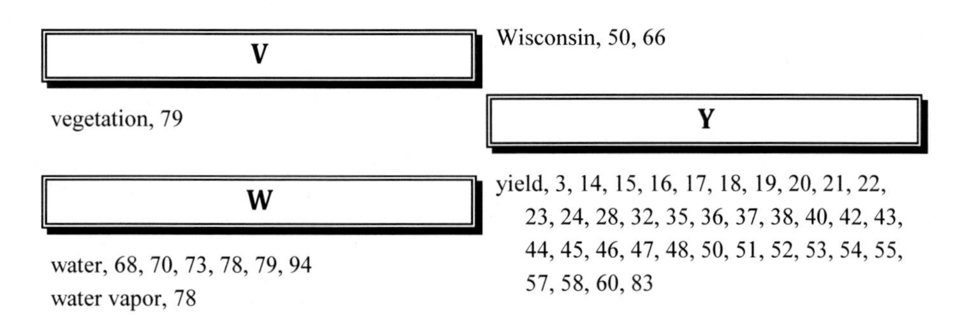

V

W

Y